Primary Science

Multilingual Matters

Child Language Disability: Volume I, II and III
    KAY MOGFORD-BEVAN and JANE SADLER (eds)
Critical Theory and Classroom Talk
    ROBERT YOUNG
Deaf-ability - Not Disability
    WENDY McCRACKEN and HILARY SUTHERLAND
Education for Work
    DAVID CORSON (ed.)
Emerging Partnerships: Current Research in Language and Literacy
    DAVID WRAY (ed.)
Equality Matters
    H. CLAIRE, J. MAYBIN and J. SWANN (eds)
Language Policy Across the Curriculum
    DAVID CORSON
Local Management of Schools
    GWEN WALLACE (ed.)
The Management of Change
    PAMELA LOMAX (ed.)
Managing Better Schools and Colleges
    PAMELA LOMAX (ed.)
Managing Staff Development in Schools
    PAMELA LOMAX (ed.)
One Europe - 100 Nations
    ROY N. PEDERSEN
Parents on Dyslexia
    S. van der STOEL (ed.)
Performance Indicators
    C. T. FITZ-GIBBON (ed.)
Policy Issues in National Assessment
    P. BROADFOOT *et al.* (eds)
Psychology, Spelling and Education
    C. STERLING and C. ROBSON (eds)
School to Work in Transition in Japan
    KAORI OKANO
Story as Vehicle
    EDIE GARVIE
Teacher Supply and Teacher Quality
    GERALD GRACE and MARTIN LAWN (eds)

Please contact us for the latest book information:
Multilingual Matters Ltd,
Frankfurt Lodge, Clevedon Hall, Victoria Road,
Clevedon, Avon BS21 7SJ, England

Primary Science

The Challenge of the 1990s

Edited by

Lynn D. Newton

MULTILINGUAL MATTERS LTD
Clevedon • Philadelphia • Adelaide

Library of Congress Cataloging in Publication Data

Primary Science: The Challenge of the 1990s/Edited by Lynn D. Newton
p. cm.
Also published as a special issue of Evaluation and Research in Education, v.6,
no. 2-3, 1992
Includes bibliographical references
1. Science–Study and teaching (Elementary)–Great Britain.
I. Newton, Lynn, D., 1953- . II. Evaluation and Research in Education, v. 6.
LB1585.5.G7P75   1992
372.3'5'0941–dc20

British Library Cataloguing in Publication Data

A CIP catalogue record for this book is available from the British Library.

ISBN 1-85359-176-9 (hbk)

Multilingual Matters Ltd

UK: Frankfurt Lodge, Clevedon Hall, Victoria Road, Clevedon, Avon BS21 7SJ.
USA: 1900 Frost Road, Suite 101, Bristol, PA 19007, USA.
Australia: P.O. Box 6025, 83 Gilles Street, Adelaide, SA 5000, Australia.

Typeset by Photographics, Honiton, Devon.
Printed and bound in Great Britain by Short Run Press Ltd, Exeter.

Contents

PRIMARY SCIENCE: THE CHALLENGE
OF THE 1990s

Lynn D. Newton

*School of Education, St Thomas' Street, University of Newcastle upon Tyne,
NE1 7RU*

Abstract   The last decade has seen the establishment of science as a core subject in the curriculum of primary schools in England and Wales. The focus of research and development has tended to be on the needs of the children and the development of a curriculum which will meet those needs. But what of the teachers themselves? Are they ready to meet the demands generated by these rapid changes? This special issue of *Evaluation and Research in Education* considers this problem, bringing together some research and experiences of science educators from the United Kingdom and abroad, not only highlighting the problems but also suggesting possible ways forward.

In the United Kingdom we are living at a time of tremendous educational change. The economic growth and optimism of the 1960s and early 1970s have been replaced in the late 1970s and 1980s by falling rolls, limited resources and devolved management in education. This change culminated in the Education Reform Act of 1988, which brought sharply into focus issues related to policy and management, curriculum structures, assessment procedures and performance indicators. As Keith Evans, Director of Education for Clwyd, stated:

> It is a truism of our time that 'the only certainty is change' and this has been borne out most dramatically in recent years in the world of education. (Evans, 1989)

The consequences of the Education Reform Act Pollard (1990) describes as 'the biggest challenge this century' facing those of us involved in education.

> . . .the 1990s will be a decade in which we will all have the opportunity to test the propositions and goals that are now enshrined in legislation. (Pollard, 1990)

The National Union of Teachers, in its recent paper in Primary Issues, states that primary teachers have borne the brunt of the changes which have resulted from the imposed Government reform. The paper suggests:

> Reforms imposed by the Government since the mid 1980s have been characterised by a lack of consultation with teachers, unacceptable pace and a total disregard for the need for resources to support their implementation. (NUT, 1992)

It is also true that:

> The task of the primary teacher has changed dramatically in recent years, particularly since the arrival of the National Curriculum. (Alexander *et al.*, 1992)

Primary education in England and Wales has been the focus of much debate over the past twelve months. It has culminated in the publication of the report by Alexander, Rose & Whitehead (1992), instigated by the Secretary of State for Education because of his belief that standards in primary education have fallen. It is hoped that action resulting from the evidence and suggestions in the report will improve standards. However, Pollard (1990) considers that this search for improved standards of education is comparable to that for the Holy Grail. Despite the pressures on primary teachers outlined in the NUT paper, the HMI report on the response to the Education Reform Act (HMI, 1991) indicates that there has been progress and improvement generally in terms of quality of learning, parental access and planning for match. According to Pollard (1990), the challenge of the 1990s will be to ensure that the new structures evolving from the Education Reform Act are a success from the perspective of the children in our care.

In primary education, one of the greatest areas of change over the past two decades has been in the teaching and learning of science. The lack of provision of good quality science education for young children had been highlighted in the publication of the conclusions of the HMI survey, *Primary Education in England*:

> . . .the progress of science teaching in primary schools has been disappointing; the ideas and materials produced by curriculum projects have had little impact in the majority of schools. (DES, 1978)

Lack of guidelines or written schemes of work, lack of appropriate equipment and resources, inadequate development of scientific skills and coverage of appropriate knowledge and understanding and lack of match of experiences to abilities were all commented upon in the report. The report also recommended better deployment of existing science specialists in primary schools and In-Service training. The 1988 Act has defined the National Curriculum and identified science as a 'core' subject for all pupils beside mathematics and English. By the 1990s investigators are finding that things have changed dramatically as far as primary science is concerned.

> . . .there have been improvements in the quality of teaching in, for example, science. . . (Alexander *et al.*, 1992)

Putting aside the current confusion over the old attainment targets (OATs) and the new ones (NATs) in the National Curriculum for Science (DES, 1991), all primary schools should now have clearly co-ordinated schemes or programmes for science. With the help of the non-statutory guidance such programmes should build upon principles of good practice in science education like those described in the 1985 policy statement (DES, 1985). Thus, the experiences offered to children should be broad and balanced, covering both the scientific skills and processes and science as a body of knowledge and also physical as well as natural aspects of science. Experiences should be relevant, and matched to each individual child's needs and abilities. Resources should now be accessible to all teachers to facilitate these science experiences, both in terms of commercially produced schemes and materials and equipment for scientific investigation. The in-service initiatives of the 1980s, particularly the educational support grant programme which began in 1984 (and in some local authorities is still continuing) should mean that primary science coordinators can offer specialist support throughout primary schools.

So, is this reality? The collection of papers brought together in this special issue looks at the current position of primary science. The message which comes through in all of the papers is that all is not well with primary science. A national survey of primary teachers carried out by Wragg, Bennett & Carre in 1989 had found that science was one of the three subjects that teachers felt themselves to be least competent to teach (Wragg *et al.*, 1989). The curriculum is in place. The children are set to go, but many primary teachers are far from ready because they lack the knowledge base necessary to deliver National Curriculum science effectively. This problem had been recognised by HMI in 1978, when they identified the teachers' own lack of knowledge of elementary science as the most severe obstacle to effective primary science teaching (DES, 1978). The concern was reiterated in 1990 by Bennett:

> Most serious however is the ability of teachers to deliver effectively the new curriculum. Of particular concern must be teachers lack of subject matter knowledge in areas like science. . . (Bennett, 1990)

The NUT paper (1992) also expresses concern about the 'very considerable' knowledge demand on teachers, especially towards the end of Key Stage 2. In 1988, McDiarmid *et al.* had cited evidence that teachers' subject knowledge is crucial if they are to be effective in making curricular decisions, selecting tasks, posing questions and evaluating pupils' understanding:

> . . .the actual subject knowledge of many primary teachers is a little shaky in some areas, such as science, and this inevitably causes difficulties in interpreting the challenge of learning tasks and in diagnosing each child's needs. It is clear that support and help for teachers as they develop their professional expertise must be provided. (Pollard, 1990)

Few in education would disagree that teachers' subject knowledge is important, and therefore some action must be taken to enhance it, particularly in the context of science. In the Association for Science Education statement, *Change in Our Future* (ASE, 1992), eight priorities are identified for the professional development of teachers of science. One is that teachers need to be confident of their own personal knowledge:

> The curriculum can only be what the teacher and learners make of it . . . Teachers rely on a knowledge and understanding of the learning process and not on the learning matter, although a teacher's knowledge about what is taught is usually a pre-requisite. (ASE, 1992)

One suggestion of Alexander *et al.* (1992), and one which had been made in the 1985 policy statement (DES, 1985), is the use of subject-specialist teachers, particularly towards the end of Key Stage 2. Given that the majority (over 80%) of primary schools have insufficient teachers to have a specialist in every curriculum area, this implies that some primary teachers will have to be specialists in more than one area. To what extent is this possible? The NUT paper (1992) suggests that this would still not solve the problem of lack of subject-knowledge among the other teachers. It would be better to have reduced class sizes, a point also endorsed in the Association for Science Education Policy Statement (ASE, 1992). ASE suggests that for all pupils from 5 to 16, where practical work is concerned, class sizes of not more than 20 are desirable. This would facilitate a better quality of teaching and learning, better pupil–teacher interaction and better opportunities for formative and summative assessment.

The National Curriculum for Science focuses on two facets of science: science as a way of thinking and working, the processes of science, and science as a body of knowledge, the products of science. Each aspect is given equal weighting. Douglas Newton, in his paper *Children Doing Science: Observation, Investigation and the National Curriculum for England and Wales* describes evidence that primary teachers lack awareness of the nature and function of different science activities in extending children's scientific process skills and competencies. He suggests that because primary science is often taught by teachers with little or no expertise in the subject and experience of science as a process, they lack understanding of the process skills involved. The activities they provide develop mainly observation and communication skills. He also points to evidence that suggests that the use of published support materials helps teachers provide activities which allow a wider range of skills to be practised. This is of practical importance, since it suggests one readily available solution to the problem.

The emphasis on investigative skills as well as knowledge and understanding in science is also discussed by Alan Peacock in his paper *What Can We Learn from Other Countries About Teaching Science Investigation Skills to Primary Children?*. He considers that practice in primary science is a long way from the ideals of the National Curriculum. Given that the cost of attempting to change primary teachers' practice to include process science is huge in terms of time, resource provision and in-service support, he asks if it is worthwhile. He compares approaches to primary science teaching in a number of other European countries and also in some African countries. He concludes by raising important questions about teaching and learning in primary science and asks whether science education at this stage is as important as is often assumed.

In her paper *Primary Pupils Conceptions of Energy: Fact or Fiction?* Gill Nicholls describes research into children's prior conceptions and alternative frameworks associated with concepts of energy. She suggests that teachers need to take into account such prior conceptions when developing teaching materials. She discusses possible underlying structures in the way pupils view energy, looking particularly at the construction of new concepts on the basis of current knowledge. Given that the teachers need to have a good understanding of the knowledge children have prior to scientific experiences, it follows that they themselves must have an appropriate understanding. It is also essential if quality teaching materials are to be developed to aid skill development and concept aquisition.

Mike Summers and Colin Kruger take up this point in their paper *Research into English Primary Teachers' Understanding of the Concept 'Energy'* describing a study of teachers' own concepts of energy. Their research was conducted in response to the concern which was being expressed about the ability of primary teachers to meet the demands of the National Curriculum for science. They discuss the implications of their findings that primary teachers lacked knowledge and understanding of the concept of energy and raise questions about the general level of conceptual understanding of science possessed by the teachers themselves.

In *Tackling Contradictions in Teachers' Understanding of Gravity and Air Resistance*, Robin Smith and Graham Peacock echo Summers and Kruger, expressing concern over teachers' knowledge and understanding of gravity and air resistance, and they suggest that many primary teachers lack the necessary scientific knowledge to deliver the science curriculum.

The changing expectations made of primary teachers in respect of their science subject-matter knowledge and understanding over the past two or three decades are reviewed by Terry Russell and his colleagues in *Teachers' Conceptual Understanding in Science: Needs and Possibilities in the Primary Phase*. They suggest that the evidence points to the advantage of enhanced subject-matter knowledge but suggest that knowledge does not necessarily equip teachers to teach primary science. Through a review of various projects carried out by the Centre for Research in Primary Science and Technology, they discuss how best to support teachers and enhance the knowledge base.

In their paper on primary practice, Alexander *et al.* (DES, 1992) emphasise the importance of questioning as a teaching and learning technique. They also discuss teachers' inabilities to use questioning effectively. In her paper, *Raising and Answering Questions in Primary Science: Some Considerations*, Catherine Woodward reviews the literature on questioning in primary science, with reference to the requirement in the Statutory Orders for the National Curriculum for Science that teachers should encourage children's questioning skills. She identifies the apparent difficulties which teachers have in implementing this requirement because of their lack of knowledge and understanding of science and their lack of skill at using questioning techniques in interrogative contexts. She considers that teachers do not view questioning as having pedagogical value and suggests that teachers need to be supported in their professional development to review their pedagogical beliefs and develop questioning technique.

The economic recession has resulted in financial cuts in education, one result being a decline in support from Local Authority advisory teachers for primary science and the closure of specialist LEA funded centres. This throws the schools on their own resources and makes the effective science co-ordinator a valuable asset. The complexity of the role of the science co-ordinator in primary schools is discussed by Derek Bell in *Co-ordinating Science in Primary Schools: A Role Model*. He suggests that a view of the role restricted to the production of policy and organisation and management of resources is inadequate if the demands of the National Curriculum for science are to be met. Experience, expertise and flexibility to adapt to varied situations are all qualities necessary for the co-ordinator to support good science education throughout the school. HMI (1991) had found that INSET training of co-ordinators was not enough—dissemination in schools was crucial. All too often INSET influences practice only in the classrooms of those who experience the INSET. If a co-ordinator is to have real effect, a planned and flexible programme for working beside colleagues is essential.

In *Science in Primary Schools: Innovation and Effectiveness in Northern Ireland*, Frank Curragh and Patricia Diamond identify and classify the indicators of effectiveness which govern the implementation of a programme of primary science education. Of the six efficacy measures, the professional competence of the teachers was found to be of prime importance. This included resourcing and the leadership potential of individuals. Other considerations included teaching and learning methodologies and the psychological set of the school.

The situation in New Zealand regarding primary science is very similar to the United Kingdom. Because of teachers' lack of confidence and competence in science, little or no primary science was being taught and where it did occur misconcepts were being formed. In their paper, *Developments in Primary Science: A New Zealand Perspective*, Fred Biddulph and Malcolm Carr describe a major research project seeking to overcome the difficulties encountered. As they succinctly state:

> It is one thing to propose an acceptable goal for primary science education; it is quite another for teachers to translate this into practice.

This collection of papers shows that there is a need to monitor the progress of primary science over the remainder of the decade. It is still in its infancy and it is probably too early to make definitive statements about its final quality. What we do know is that some teachers *are* teaching primary science and teaching it well (HMI, 1991). Quality resources are becoming available. Good practice will be supported through the use of these resources providing that they really are of high quality. They must support and encourage broad, balanced and relevant science, a range of process skill development and use of techniques like teacher *and* pupil questioning. Resources for practical work must also be of quality so that primary science can shed its yoghurt pot and pop bottle image. The co-ordinator's role is vital here, for policy development, resource management and colleague support. Some co-ordinators *are* co-ordinating science effectively throughout their schools (and sometimes across several small schools), supporting and encouraging their colleagues in a positive way. As experienced teachers gain confidence and competence in delivering the National Curriculum for Science, whether through the support of a co-ordinator, an advisory teacher, a commercial scheme or an INSET course, the situation should improve. As new teachers begin their training with a background experience of GCSE Science, which includes science as a process, and receive initial training in the delivery of the National Curriculum for Science, the situation should improve (Newton, 1991). Attitudes to science develop early and, once established, are extremely difficult to change (Newton & Newton, 1992). As the children themselves experience a broad, balanced, relevant science education they should develop positive attitudes towards science, and the situation should improve. As far as science education is concerned, this is the challenge of the 1990s—to bring about such improvements.

References

Association for Science Education (1992) *ASE Policy: Present and Future*. Hatfield: ASE.

——(1992) *Change in our Future: A Challenge for Science Education*. Hatfield: ASE.

Bennett, N. (1990) The Primary Curriculum. In T. Brown and K. Morrison (1990) *The Curriculum Handbook*. Harlow: Longman Group UK Ltd.

Department of Education and Science (1978) *Primary Education in England: A Survey by Her Majesty's Inspectors of Schools*. London: HMSO.

——(1985) *Science 5–16: A Statement of Policy*. London: HMSO.

——(1991) *Science in the National Curriculum (1991)*. London: HMSO.

——(1992) *Curriculum Organisation and Classroom Practice in Primary Schools*. A Discussion Paper by R. Alexander, J. Rose and C. Woodhead. London HMSO.

Evans, K. (1989) *Into the Future: Education in Clwyd after the Education Reform Act 1988*. Clwyd County Council: Education Department.

Her Majesty's Inspectorate (1991) *The Implementation of the Curricular Requirements of the Education Reform Act: Science Key Stages 1 and 3*. A Report of Her Majesty's Inspectorate on the first year, 1989–90. London: HMSO.

——(1991) *A Survey of INSET in the National Curriculum Core Subjects 1990–1991*. Report of Her Majesty's Inspectors of Schools. London: HMSO.

McDiarmid, G. W., Ball, D. L. and Anderson, C. W. (1988) Why staying one chapter ahead doesn't really work: Subject specific pedagogy. In M. C. Reynolds *Knowledge Base for the Beginning Teacher*. Oxford: Pergamon Press.

National Union of Teachers (1992) *Primary Issues: Our Great Expectations*. London: NUT.

Newton, L. D. (1991) Who needs trained relations? *Primary Teaching Studies* 6 (1), June 1991, 62–7.

Newton, D. P. and Newton, L. D. (1992) Young children's perceptions of science and the scientist. *International Journal of Science Education* 14, 3.

Pollard, A. (1990) *Learning in Primary Schools*. London: Cassell.

Wragg, E. C., Bennett, N. and Carre, C. (1989) Primary teachers and the National Curriculum. *Research Papers in Education* 4 (3), 17–46.

CHILDREN DOING SCIENCE: OBSERVATION, INVESTIGATION AND THE NATIONAL CURRICULUM FOR ENGLAND AND WALES

Douglas P. Newton

School of Education, University of Newcastle-upon-Tyne, St Thomas' Street, Newcastle-Upon-Tyne NE1 7RU, UK

Abstract The National Curriculum requires that science be taught as both a body of knowledge and as a way of working, that is, as products and processes. In the primary school, science is often taught by teachers without extensive personal experience of the subject, particularly of those processes required by the National Curriculum. There is evidence which suggests that there is a tendency to provide activities which mainly practise observation and communication and which neglect other investigative processes. This study describes the results of a survey of the activities provided by fifty primary school teachers in the North-East of England which corroborate these earlier findings. They suggest that the cause is not simply ignorance of investigative processes associated with science. It may also involve a lack of awareness of the nature and function of science activities provided to extend awareness, facilitate understanding, enhance learning, and practise skills and processes. There was some indication that those who used published support materials as guides were more likely to provide activities which practised a wider range of processes. A solution to the problem might be to encourage the use of such materials.

Introduction

Until relatively recently, science has tended to be a second class member of the primary school curriculum in England and Wales. Various reports have highlighted past weaknesses, like the relative absence of teaching science as a way of working and the neglect of physical science (DES, 1959; Harvey, 1976). The National Curriculum elevates science to the core curriculum, making it an equal of English and mathematics. Now, pupils must be taught the science of light, heat, sound, electricity, forces and materials as well as of plants and animals. They must also have the opportunity to work scientifically and develop appropriate skills and processes. To what extent is the National Curriculum achieving its aims?

In 1973, the Derbyshire Primary Science Enquiry surveyed teachers' objectives in primary science teaching. 'To observe accurately with all the senses' was the prime objective of the overwhelming majority with 'communicate' as a close second. Other processes, like interpreting results, came low in the list of priorities (Bradley, 1976). In 1986, Skamp reported on a small survey of the opportunities provided for learning process skills in the primary school. In the ten lessons analysed, the practice of observation was the skill most often noted while other skills, like hypothesising and raising questions for investigation were least exercised. During 1989–90, soon after the introduction of the National Curriculum for science at Key Stage 1 (5 to 7 year olds), Her Majesty's Inspectors found that Infant teachers gave a high priority to the development of observation and communication skills and placed much less emphasis on other skills and processes. They added that 'in many schools insufficient emphasis was given to helping pupils progress from wild guesses to sensible suggestions informed by personal experience'. It was also found that the children's work was insufficiently linked to the subject knowledge prescribed by the National Curriculum and that there was little confidence in teaching physical and earth science (HMI, 1991). There were also, of course, 'good lessons' associated with 'relevant starting points' and where 'pupils were encouraged to define the task and plan their own work'. Almost thirty years ago, Gagné argued that other skills or processes, like hypothesising, controlling variables and interpreting data, can be practised by all (Gagné, 1965) and Daryll Evans (1987) points out that this includes young children who, for instance, can and do hypothesise. Why, then, are some processes neglected?

One reason may be a failure to distinguish between various kinds of activity. For instance, some activities serve mainly to demonstrate the existence and nature of a phenomenon, property or interaction; others are intended to investigate aspects of these and develop process skills in doing so (Harlen, 1985). Where the first is favoured it is likely that observation and communication will be practised extensively. The second will facilitate the practice of other processes. By itself, this explanation is not enough because a teacher would soon notice an absence of opportunity in other process and skill areas. He or she must believe that the requirements of the National Curriculum are being met and that the activity offers practice in these other areas. When teachers do provide activities to exercise such skills and processes, there is evidence that they overestimate the children's abilities in them (Black et al., 1984). Jungwirth & Dreyfus (1990) have more than a suspicion that teachers of many countries are themselves deficient in their awareness of scientific skills. Even when understanding of the terms is present, if hypothesising only calls to mind conflicting theories of atomic structure and not the more homely and testable hypotheses associated with, say, the cause of lengthening shadows on a sunny day then it is unlikely that activities would be structured so that youngsters could formulate and test hypotheses.

The aim of this study was to examine some activities offered by teachers to children at Key Stage 1 (5 to 7 years) and Key Stage 2 (7 to 11 years) and to see if there is any support for the explanation given above.

Method

Half way through the Autumn Term of 1991, fifty teachers, one in each of fifty primary schools in the North East of England, were interviewed and invited to describe and discuss the current or most recent science activity they had provided for their classes. Twenty-five of these had Key Stage 1 classes and twenty-five had Key Stage 2 classes and the interview was structured by a pro forma recording sheet (Appendix). The questions were intended to generate a sufficient account of the activity to allow it to be allocated to one of the following content areas, according to its main association:

Life (Plant) e.g. flowers, vegetables, fruits
Life (Animal) e.g. ourselves, seashore-animals
Materials e.g. waterproof materials, low friction materials
Force/Energy e.g. forces and their effects, light, heat, sound, magnetism, electricity
Earth/Space e.g. rocks, soils, erosion, Solar System
Other e.g. making an artefact to satisfy a need, topics not included above.

The first five categories covered most of the knowledge and understanding requirements of National Curriculum science.

The second intention was to generate a sufficient account of the teachers' introduction to the activity to classify it as being expressly related to the everyday world or offered as an exercise divorced from everyday experience. Connection with the everyday world was considered to have been made if the activity was introduced through or related to an object, situation, event or experience likely to be familiar to the children. This would include, for instance, activities which studied directly some aspect of the environment around the children, like trees, leaves, or litter, or activities which began with, led to, or examined in more detail some common experience, like the different conducting properties of metal and plastic spoons.

The third intention was to determine whether published textual materials played a direct part in the design or application of the activity and if this related to the quality of the experience provided.

The fourth intention was to classify the activities according to whether they were to demonstrate the existence and nature of a phenomenon, property or interaction or were to develop a range of process skills, or served some other function. Each activity was analysed into the sequence of experiences it provided.

The first basic type expected was one initiated by the provision of direct experience of a phenomenon or property and calling for observation with one or more senses. Variants might offer vicarious experience in addition, as in asking for recall of everyday instances, and follow the observation stage with requests for an explanation and/or various forms of communicating and recording, like describing or drawing, but the essence of the type is the sensing of the subject of study.

The second basic type expected was pupil investigation of some aspect of the subject of study, involving a wider range of process skills. Typically, this would begin with direct or vicarious experience which would be used to generate a tentative explanation, hypothesis, question or problem to investigate, followed by a practical investigation intended to test, answer or solve these and which generated results used to draw a conclusion. Observation would be involved, at least in the practical investigation, and various forms of recording and communicating could be expected. The essence of this type is the early appearance of the explanation, hypothesis, question or problem for testing and the practical investigation, followed by some consideration of the meaning of the results.

In this analysis, it is not only the existence of a process which is important, but also its position in the sequence of events. A testable, tentative explanation might be the final outcome of an activity of the first kind but would be near the beginning or precede an activity of the second kind. In practice, it is possible for the first type of activity to generate an explanation to test directly in one of the second type. Where this occurred, such activities were recorded separately. Any other types of activity which occurred were similarly recorded. The account of the Characteristics of Investigative Work in the Non-Statutory Guidance to the National Curriculum for science provides some description cf the processes to be practised. Briefly, these are:

Observing: 'by which the perceptions of objects or events are selected, interpreted and their significance judged', in short, 'noting features'.
Hypothesising: 'by which a tentative, testable explanation of an observed event is produced.
Predicting: by which an outcome of an event or action is anticipated; this may be a reasonable prediction, deriving from a tentative explanation or observed regularity, or a blind guess.
Designing an investigation: by which the way of collecting relevant information is determined.
Executing an investigation: in which the determined way is enacted.
Concluding: 'deciding what the results mean', 'interpreting and evaluating data', inferring from data.
Communicating: 'keeping a record' and 'reporting in an appropriate manner'.
(NCC, 1989)

The fifth intention was to determine what the teachers thought was the relationship between their activities and the scientific skills and processes described by the National Curriculum. They were asked to select the National Curriculum groups of intellectual and practical skills that they considered their activity to practise:

(i)   plan, hypothesise and predict
(ii)   design and carry out investigations
(iii)   interpret results and findings
(iv)   draw inferences
(v)   communicate exploratory tasks and experiments   (DES, 1989)

The teachers' views were compared with the type of activity they offered.

Results

Content area

The content area associations are shown in Figure 1. At Key Stage 1, most of the activities were associated with the natural sciences, especially plant life. For example, a Reception class (4–5 years) named and described a pineapple, melon, lemon, grapefruit, banana and parsnip. They classified them according to the roughness of their skin and tested them to see which would roll down a slope (based on a published scheme). A Year 1 class (5–6 years) examined the colour and texture of vegetables and drew pictures of them. Another Year 1 class watched the growth of mould on various foodstuffs, observing and predicting changes. In Year 2 (6–7 years), one class compared two different plants, using a range of senses, and communicated their observations with drawings. Another class examined apples and predicted what would happen when an apple was cut and left in the air.

Of the less popular content areas, some Year 1 children were given a box of different materials and had to think of different words to describe them, then sort them into groups according to properties such as texture. Year 2 children tested boats they had made, the effect of forces on toy cars, and classified materials people throw away.

At Key Stage 2, the majority of the activities were associated with the physical sciences (Figure 1). For example, a Year 3 class (7–8 years) tried ways of reducing the friction opposing the motion of a brick, another such class observed images in mirrors. Some Year 4 pupils (8–9 years) classified materials as solids or liquids and changed some from one state into the other. Other pupils of this age had to find objects which float and objects which sink then make one which sinks into one which floats and explain why some things sink. In Year 5 (9–10 years), some pupils observed the action of colour filters and saw the effect of mixing light of different colours while others tested the solubility of salt and sugar in warm and cold water. Some Year 6 (10–11 years) pupils tested the properties of different materials and hence selected the clothing best suited for taking on an outdoor school trip. Another such class felt the conduction of heat through everyday objects.

Of the less popular content areas, a Year 4 class visited the seashore and observed and identified sea-creatures; a Year 5 class watched a plant take up coloured water through its stem, and a Year 6 class examined teeth imprints and discussed the function of different kinds of teeth.

At both stages, a small number of teachers described the designing and making of an artefact as their most recent or current scientific activity (three at Key Stage 1 and two at Key Stage 2). For instance, some children designed and made a pinwheel (Year 2), some invented a game which used magnets (Year 2), and one

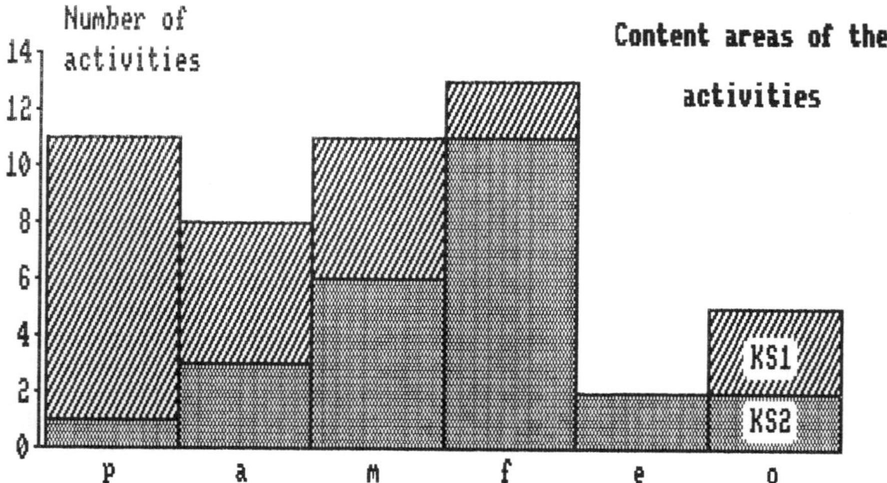

Figure 1 The number of activities classified by content area at Key Stages and 2 (p = Plant life, a = Animal life, m = materials, f = Force/Energy, e = Earth/Space, o = Other).

class made a model of the Solar System (Year 4). One Year 5 class observed the use of divining rods. This small group of activities is shown as Other in Figure 1.

Explicit real-world connection

In introducing the activity, all the activities at Key Stage 1 and 14 at Key Stage 2 were either explicitly connected to the real world, as when they had to find the best clothing for a school trip, or were a direct study of an aspect of the real world, as in a study of seashore life. The remainder were presented as self-contained exercises whose relevance was not overtly expressed at the time, as when the passage of light through coloured filters was explored.

Expressed use of published textual materials

At both Key Stage 1 and Key Stage 2, the majority of teachers claimed not to rely on a published scheme or similar textual material for the idea for the activity or to have used such material to present the activity. Only three teachers at each stage described the published source of their activity.

Type of activity

At Key Stage 1, 17 of the activities were of the first type, providing direct experience of a phenomenon, property or relationship and demonstrating its existence. Four others began with this kind of activity and developed it into the second type. One provided the accepted view and verified it by demonstration. The remaining three activities were technological in nature.

At Key Stage 2, 13 of the activities were of the first type, six were of the second type, with one other beginning with direct experience and developing it into an investigation of the second type. Two others were verifications of explanations provided at the outset, two were technological and one was completely vicarious experience.

While it is tempting to suggest that there are proportionally more of the second type of activity at Key Stage 2, the difference for a sample of this size is not statistically significant. Figure 2 a and b shows the proportions of the two main types of activity at each Key Stage.

The teachers' views of what their activity offered (for those providing practical experience in science) are shown in Table 1.

For an activity with Reception children, to do with rolling fruit down a slope, the teacher considered it provided the opportunity to plan, hypothesise and predict, interpret findings and communicate. These processes seemed to be exercised by the activity as it was described. Another activity (Year 2) required the children to describe an apple, predict what would happen to the inside when it was cut, record what they observed, then think of a way of preventing cut apples going brown. This teacher claimed all the processes listed. Another Year 2 class read a recipe for making jam, weighed and prepared the fruit, watched the jam being made and guessed what it would look like when it was finished. The teacher felt the exercise practised all the processes listed except 'design and carry out investigations'.

All activities at Key Stage 1 required the pupils to observe and describe. In four of these the pupils were expected to compare, contrast and hence classify,

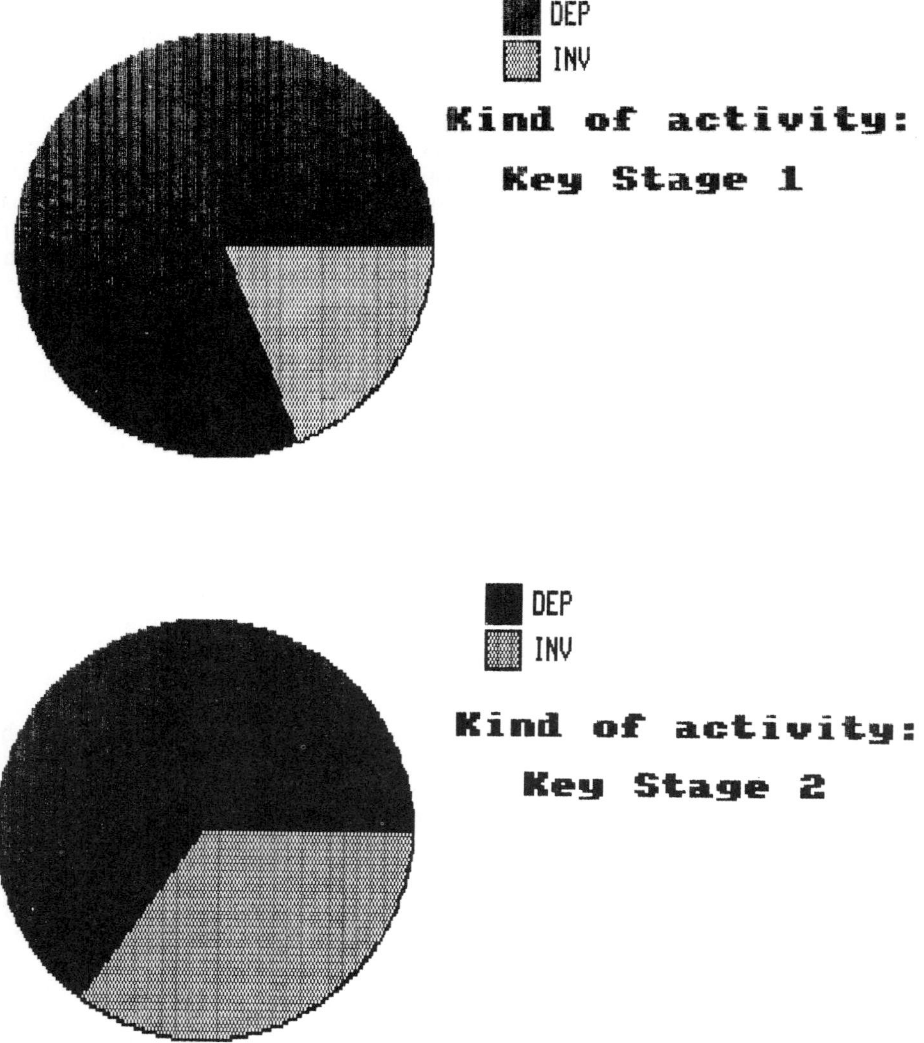

Figure 2 Proportions of activities at Key Stages 1 and 2 which were classified as providing direct experience of a phenomenon or demonstrating its existence only (DEP) or of practical investigative skills like hypothesis testing (INV).

as when examining the physical properties of materials. About a third of the activities asked for an explanation of what was observed. Two called for a 'fair test' to be made, as in testing boats of different designs. Eight asked pupils to predict what would happen next, as in 'How long will it take?' in reference to the uptake of coloured water by a plant. It seemed better to describe prediction of this kind as either simple recall or guessing. Three asked for predictions to be made in the sense of an informed or reasoned guess, as in 'Will this fruit roll?' after the children had tested others of various shapes.

Table 1 The proportion of teachers who saw their activities as practising each National Curriculum group of skills (for those providing practical experience in science)

	Key Stage 1		Key Stage 2	
	DEP	INV	DEP	INV
	$n=17$	$n=4$	$n=13$	$n=7$
Plan, hypothesise and predict	0.82	1.00	0.46	0.86
Design and carry out investigations	0.53	0.75	0.54	0.86
Interpret results and findings	0.71	1.00	0.77	1.00
Draw inferences	0.94	0.75	0.46	1.00
Communicate tasks and experiments	0.76	0.75	0.54	1.00

A Year 6 teacher described an activity to do with finding the best materials for a tent. This was presented in a fairly open way with the children expected to decide what variables were important, what they would do, carry out the investigation, arrive at a conclusion and recommendation and communicate their findings. The teacher felt that, although some children needed guidance, this activity provided the opportunity to practise all the processes listed above. A Year 4 activity required the children to draw a pattern, examine its image in a mirror, draw what they saw and note how the image and object differed. This teacher claimed all processes were practised except the drawing of inferences. Another Year 4 activity required children to design and make a Solar System mobile from paper, card and string, with the help of reference books. The scientific processes claimed were planning, designing and investigating.

At Key Stage 2, all activities again required pupils to observe and describe and about a third asked for an explanation. Classification figured in two of them and seven called for a 'fair test', as in testing materials to make a suitable tent. Five asked for recall/guess predictions, as when the uptake of coloured water by a plant gave rise to the teacher's question, 'Will it kill the plant?' Two asked for informed or reasoned guess predictions, as when children had to predict how the force needed to move a brick on rollers would compare with what they had found it was on a carpet.

Discussion

The Key Stage 1 study of 1989–90 by HMI had detected an emphasis on observation and communication skills and a tendency to accept guessing in place of reasoned prediction. These were also noted here. HMI also found there was little confidence in teaching physical and earth sciences. The study reported here was conducted in the Autumn Term. Autumn provides opportunities for work on such things as seeds and wild fruits so a significant number of plant and animal activities deriving from the season might be expected. In practice, this was not the case. At Key Stage 1, there was an activity to do with fruit brought in for the harvest festival and one on jam making; the others could have been done at almost any time of the year. Therefore, the pre-dominance of natural science activities at this stage did not seem to be due to the season. It could arise from teachers' lack of confidence in other areas of science so that associated activities are perceived less comfortably, or less readily, or as being more difficult and best postponed until later in the school year. Taken together, these results are in accord with those of the earlier study and give some credence to the findings.

At Key Stage 2, the pattern of content areas is reversed and physical science dominates while natural science takes a back seat. Teachers of juniors (7–11 years) do not seem to share their infant (4–7 years) teaching colleagues hesitancy about aspects of physical science like force and energy. Historically, some kinds of natural science have been seen as appropriate to the primary school and a body of well-developed practice is available to the teacher. Physical science teaching, seen through the eyes of secondary school experience, was probably excluded more from infant practice than from junior so infant teachers may feel more remote from it and hence less confident about teaching it. Such perceptions are likely to give physical science at this level an aura of difficulty which could be reflected in teaching sequences; schemes begin with natural science and progress to physical science. While both areas of science are required by the National Curriculum at both Key Stages, there may be a tendency for a progression from natural to physical science within each stage and, perhaps, between the stages.

An emphasis on observation at both Key Stages seems likely to be associated with a tendency to favour activities for providing direct experience of a phenomenon. The teachers who provided these in this sample commonly indicated that their activities exercised such investigative skills and processes as planning, hypothesising and predicting, and designing and carrying out investigations. This suggests that there is some misunderstanding about the nature of these skills and processes. Although this seems to be true at both Stages, Infant teachers seem even more inclined to be immoderate in their claims than Junior teachers, their distribution of endorsements of some categories being significantly greater than those at Key Stage 2 (for example; plan, hypothesise and predict, $p<0.05$, Fisher test). Other evidence of a vagueness in understanding of the processes may be a common confusion of blind guessing with informed prediction, mentioned earlier. There are also those who described the making of an artefact as their scientific activity and claimed the practice of such processes as the design and execution of an experiment. In essence, these seemed to confuse designing and making as defined in National Curriculum technology with the scientific processes. It should be added that this did not seem to apply in all cases and, especially where the use of published textual materials was acknowledged, some teachers seemed able to identify the National Curriculum processes practised by their activity. At the same time, this is not to suggest that the activities were, in themselves, worthless. As teaching devices, many seem valuable but it is essential that their function is recognised.

While teachers who are science graduates may fail to point out the relevance of science in the real world (Newton, 1988), others may lack the confidence and the breadth of science understanding to relate what they teach, not only to the real world, but to the everyday world of the child. This did not seem to be a problem at Key Stage 1 but at Key Stage 2 a significant number of teachers did not introduce the activity from a familiar object or relate it to everyday experience. In such cases, there was a tendency to provide the activity as an exercise which stood alone, unconnected with previous experience although it is possible that this was rectified at a later time.

Progression is also needed, of course, within any area of science. Here, activities were sampled only once in each school so progress within a school cannot be commented upon directly. However, it was noted that an activity deemed appropriate with a class in one school was sometimes found in a similar, or even less demanding form, in another school with older children. For instance, a Reception class in one school and a Year 2 class in another examined the interior and exterior of apples to much the same end. Similarly, a Year 2 class at Key Stage 1 and a Year 4 class at Key Stage 2 did much the same activity with boats. While it might be argued that the quality of the outcome was different, there seemed little difference in teacher expectations. Since these activities seemed to be successful with younger children then it may be that the teachers of older children were expecting too little of them.

In many cases, the prime function of an activity seemed to be to introduce, highlight the existence of, or focus attention upon some phenomenon, regularity or property. From such an awareness raising exercise might come the curiosity and questions which lead to an hypothesis, an experiment, and a conclusion. A visit to the sea-shore, an examination of teeth marks, the feel of heat conducted through a spoon, all provide opportunities for observing, classifying, recording and communicating. But they are also intended to stimulate curiosity and provide food for thought. Why, for instance, are some teeth marks narrow and sharp, others round and deep, and still others broad and irregular? This kind of activity often ends by asking the children to explain what they saw. Perhaps they come up with the accepted view or perhaps they do not; there the matter ends. The introductory experience, essential as it is, often does not lend itself to the practice of the range of processes required by the National Curriculum. Any general failure to turn such activities to this end would account for the dominance of observing and communicating noted here and elsewhere. There is a need to distinguish between different kinds of practical activity according to their function and to be aware that, while each has an important role, they do not do the same thing. A consequence of a lack of confidence, a confusion over the meaning of processes and a failure to recognise the function of an activity and its limitations is that pupils may not be practising skills and processes as defined by the National Curriculum and which the teacher will, presumably, claim in their assessment of the pupils' progress.

Conclusion

There is evidence that science as a way of working and as defined by the National Curriculum, is not well-understood amongst some primary school teachers. This seems to be compounded by a lack of awareness of the functions of the practical activity. While activities which illustrate the existence of a phenomenon are common, opportunities are not always taken to develop these into question raising or hypothesis generating activities for further investigation. While teachers may believe that their activity exercises a wide range of skills and processes, in practice, they seem to be over-optimistic.

There were indications that, where published support materials were used, activities exercised a wider range of skills than those practised in the self-devised, phenomenon-demonstrating activities. In-service support is often intended to provide practising teachers with some knowledge of science and its processes. By their short-term nature, it is often not easy to provide more than a smattering of knowledge and a few exemplars of good practice. It might be more effective if this was tied more closely to the published schemes and similar teaching materials available to the teacher. The aim would be to ensure that the teacher could see what the material was trying to do, why it was doing it, how to use it, recognise its strengths and weaknesses, and supplement or adapt it as necessary. Many schools are in a position to do this for themselves or with the support of an adviser. 'What we have most to wish for,' said Matthew Arnold, 'is the guidance of a good textbook' (Silver, 1977) and for many teachers, a good textbook is their main source of in-service training.

Appendix: Children Doing Science

Circle the year(s) this relates to: **R 1 2 3 4 5 6**

1. Regarding the most recent practical activity in science provided for the children:

(a)   Describe the activity (as seen by the teacher)

(b)   State the children's brief (i.e. the activity as it was presented to the children, the words used to introduce the task)

(d)   State any questions asked of the children relating to the activity

(e)   If the activity is taken from or based upon a published book/scheme, what is it?

Title: . . . .   Page/Card: . . . .

2. Regarding the process of science exercised by the activity:

Tick the processes the teacher considers that it practised on the list below:

. . . (i)   plan, hypothesise and predict
. . . (ii)   design and carry out investigations
. . . (iii) interpret results and findings
. . . (iv)   draw inferences
. . . (v)   communicate exploratory tasks and experiments.

References

Black, P., Harlen, W. and Orgee, T. (1984) *Standards of Performance—Expectations and Reality (Assessment of Performance Unit Paper No. 3)*. London: DES.

Bradley, H. (1976) *A Survey of Science Teaching in Primary Schools*. Nottingham: University of Nottingham.

Daryll Evans, J. (1987) Encouraging science hypotheses in the primary school. *School Science Review* 68, 358–61.

DES (1959) *Trends in Education 12: Young Children and Science*. London: HMSO.

——(1989) *Science in the National Curriculum*. London: HMSO.

Gagné, R. M. (1965) *The Psychological Basis of Science—A Process Approach. AAAS Miscellaneous Publication 65–68*. Washington: AAAS.

Harlen, W. (1985) *Teaching and Learning Primary Science*. London: Harper and Row.

Harvey, T. J. (1976) Science in the primary school—an estimate of the amount taught and the teacher's attitude towards it. *School Science Review* 57, 770–4.

HMI (1991) *The Implementation of the Curricular Requirements of the Education Reform Act: Science, Key Stages 1 and 3. A Report by HM Inspectorate on the First Year, 1989–90*. London: DES, HMSO.

Jungwirth, E. and Dreyfus, A. (1990) Diagnosing the attainment of basic enquiry skills: The 100-year old quest for critical thinking. *Journal of Biological Education* 24 (1), 42–9.
National Curriculum Council (1989) *Science Non-Statutory Guidance*. York: NCC.
Newton, D. P. (1988) *Making Science Education Relevant*. London: Kogan Page.
Skamp, K. R. (1986) Investigating opportunities for learning science. *Primary Science Review* 2, 10–11.
Silver, H. (1977) Nothing but the present or nothing but the past? (Inaugural Lecture, 17th. May).

WHAT CAN WE LEARN FROM OTHER COUNTRIES ABOUT TEACHING SCIENCE INVESTIGATION SKILLS TO PRIMARY CHILDREN?

Alan Peacock

University of Exeter, School of Education, St Luke's, Heavitree Road, Exeter EX1 2LU, UK

Abstract The paper examines the explicit and implicit rationales, in several countries, for the place of science in the primary school curriculum. It relates these to notable contrasts between such countries as UK and Kenya on the one hand, which emphasise the teaching of science investigation skills; and Germany and France on the other hand, which stress the acquisition of knowledge. This comparison is used to highlight questions about the impact of current primary science policy in the English National Curriculum; about assumed links between science teaching and economic progress; and about the ambitions of our broader view of primary schooling in England. It concludes by proposing further international study of the actual learning of investigation skills in the primary phase.

Introduction

Emphasising investigative skills as well as knowledge in primary science has become and remained an accepted principle within the National Curriculum, despite several revisions of the document and despite the increasingly critical reviews of 'process science' such as those by Millar & Driver (1987) and more recently Claxton (1991). It is also clear that practice in primary schools is still a long way from matching up to the ideals of the National Curriculum in this respect, even though the emphasis on science as a core subject has increased, and most teachers now include something called science in their schemes (ASE, 1988; Kinder & Harland, 1991). At the same time, the cost of attempting to modify primary teachers' practice to incorporate 'process science' has been huge, in terms of curriculum development, resource provision and INSET. So how worthwhile has this attempted transformation been?

This article does not attempt to cover similar ground to that covered by the comprehensive reviews and evaluations referred to above, by looking into what teachers do or what children learn. Instead, it tries to compare approaches to primary school science teaching in a number of other countries of Europe and Africa with our own, briefly considering their rationales in relation to national aspirations, and what practice looks like as a consequence. From these comparisons, I have tried to raise what seem to be important questions about the teaching and learning of what the National Curriculum now chooses to call Investigation Skills; and in passing, to reflect briefly on whether science education for primary age children is as important to national and economic development as is still often assumed and claimed.

Science Education and Economic Development

When I was choosing which 'A' levels to study over 30 years ago, I was persuaded by parents and teachers alike to do science subjects, even though my best 'O' level results were in maths and languages, because (they all told me) that was where the future lay. It was going to be a world in which science would be crucial: good jobs, status, success and the country's future would all depend on more and more of us studying science.

That idea has been part of the orthodoxy in education and in economic policy ever since. Harold Wilson talked about the 'white heat of the technological revolution' and put this thinking at the heart of his industrial and educational strategy. Science education was expanded as the key means of achieving these goals, curriculum development exploded in the form of Nuffield Science and other competing schemes. By 1978 the issue had moved into the primary sector as a consequence of inspectorate warnings about how bad primary science was. The HMI Primary Survey (DES, 1978) pointed out that few primary schools had effective programmes for teaching science; that in very few classes were science skills taught; and that the most severe obstacle was teachers' own lack of knowledge of elementary science, despite the existence by then of comprehensive primary science schemes such as the Schools Councils 'Science 5–13' programme, begun in the early 1970s.

Having joined the EEC, increasing British awareness of the economic success of other European countries reinforced ideas about the connection between economic progress and scientific/technological literacy: more and better science education was required. The more that successive governments failed to catch up with France and Germany in economic terms, the more science was emphasised: until in 1985 the DES took the unprecedented step of publishing a Policy Statement on 'Science For All' which stressed the case for devoting resources to science in terms of:

> the importance of science education to the pupil and to society as a whole. Science and technology permeate almost every aspect of daily life. Each of us needs to be able to bring a scientific approach to bear on the practical, social, economic and political issues of modern life. (DES, 1985)

The advent of the National Curriculum then gave science the accolade of 'core' status, and with it a major share of the curriculum cake. The Science Working Group's reports continued to stress the same theme of the importance of science in 'changing the nature of society and our economy', and resources such as ESG money became available to provide in-service training in science for primary teachers. Nobody argued about the essential provision of science, as both a methodology as well as a body of knowledge, for primary children from the age of five upwards. And throughout the 1980s, primary science INSET at both local and national level tended to focus on both these aspects of the subject. Studies did suggest (Peacock, 1989) that some parents were puzzled and perhaps a little worried at the prospect of their five-year-old doing science; would (s)he be safe (with all those test-tubes and bunsen burners!)? But in contrast to some of the other subject panels and working groups whose documents appeared later, the Science Working Group's emphasis on 'Exploration Skills' being equal in importance to science knowledge for young children was accepted at the time by teachers, scientists and the Secretary of State with little controversy.

Primary Science in Germany and France

It would be easy to assume from all this, therefore, that those countries whose economies are relatively more successful have over the years done a better job of teaching scientific literacy to their populations, and have done this by starting children at an early age on a diet of process science, concept learning and technological problem-solving. In 1990 I visited primary schools and teacher training institutions in both France and Germany, and made a comparison of the approaches to science in primary schools in these countries, with the approach currently adopted in England and Wales, looking particularly at the differences in teaching investigative skills and the impact these might have.

In both countries, I paid particular attention to what teachers and schools emphasised in primary science, to see how these might relate to their future effectiveness as citizens in a science-oriented society. To my surprise, I found that in the schools I visited they did practically nothing which a primary science adviser from Britain would recognise as science. Nor were there many science and technology resources available: none of the schools I visited in Germany had computers in classrooms, for example. Most schools exhibited science materials—rock samples, chemical products, textiles, seeds, for example—all donated by large industrial enterprises. But these displays were usually in locked cases in corridors, and clearly not for use by children. So what part does science actually play in the German primary school curriculum?

In a German primary school or 'grundschule', science does not exist as a curriculum subject in its own right: in most regions of Germany, children in grundschule study language, maths and 'sachkunde' or general knowledge, which includes science knowledge, but which largely excludes practical, investigative work. Sachkunde has evolved from what was earlier known as 'heimatkunde', a subject established in the curriculum of the newly created German elementary schools in 1921. Heimatkunde is difficult to translate; a recent study of science education in Germany explains it as follows:

> 'Heimat' is one's native land but moreso one's native place, town or village . . . an immediate surrounding, easy to survey, untouched, unspoilt, harmonious—a home creating safety. 'Kunde' means knowledge, a knowledge based on facts, not necessarily empirical facts, but often something that is 'generally known', traditional. But Kunde also implies a teaching aspect, as its second meaning is a proclamation, announcement or publication, all three of which are one-way roads, because there is somebody who knows something and proclaims, announces it or publishes it for others. (Krucker, 1987:87)

Surprisingly perhaps, heimatkunde remained largely unaffected by the Nazi period and the post-war reconstruction of the federal republic. Only in the late 1960s was heimatkunde largely replaced by a subject called 'sachunterricht' (literally 'lessons about things'), to enlarge and redirect the scientific study of the environment. It was meant to:

encompass all objects, phenomena and events which children encounter today, including those of technology, television programmes and travel experiences. Moreover, the pupils were to be acquainted with the complex and highly-structured physical and social world of an industrial society, reflecting its technical, scientific, economic, judicial and political dimensions. Emphasis was placed on science and technology as moving forces for industrial progress . . . (Lauterbach & Frey, 1987:20).

Sachunterricht was not intended as a separate science subject, nor as a set of individual disciplines. It did however require practical study, and on the face of it, looks like the kind of grounding which would serve exactly the purposes aspired to in our own society. Nevertheless, the fact is that sachunterricht never fully established itself within the curriculum. As Lauterbach & Frey go on to point out, more than a third of German states have now abandoned sachunterricht and gone back to 'sach-und heimatkunde'.

This reversion to didactic approaches is also evident in the commercially published materials which are currently used in German primary schools. For example, one 'sachbuch' (sachkunde textbook) for grade 2 (7–8 year olds) deals with 71 separate topics, of which 32 have a science-knowledge orientation, but only 8 of these topics have any active investigative dimension, as the following breakdown illustrates (numbers of topics requiring active investigation are given in brackets):

Living Things:	Ourselves	4 topics	(0)
	Plants	4	(1)
	Animals:	9	(1)
Materials:	Mixing	1	(1)
Light/Sound:	Shadows	2	(1)
	Sound	1	(0)
Structures:	Buildings	3	(1)
Environment:	Weather	2	(0)
	Earth/Space	1	(1)
Magnets:		2	(1)
Thermometers:		3	(1)

(Cornelson-Velhagen & Klasing, 1986)

It is clear therefore from watching lessons and looking at textbooks that science teaching in German primary schools is not concerned except peripherally with developing process science skills from an early age. The standard teaching strategy is one in which the teacher/textbook raises questions and then goes on to answer them for the pupils by providing the required knowledge. The pupils' role is mainly to answer questions and to learn the received knowledge.

The situation is very similar in French primary schools, where again the emphasis is on language and maths but not science. For example, whilst primary teachers during their training take a science course which is in content terms not much different from the knowledge dimension of our own national curriculum, the course does not stress open-ended investigation or the need to develop these skills in children, but rather the verifying and communicating of science knowledge and principles. A primary teacher trainee's experience of science in France would be almost entirely what we would call 'main subject' science, with little or no 'curriculum' science. In the classes I visited, the children's school week contained nothing recognisable as science.

The conclusions drawn from this limited experience are confirmed by Patricia Broadfoot's studies of the differences between French and English primary teachers (Broadfoot & Osborn, 1990), which show that French primary teachers stress product rather than process; encourage children to work on their own rather than in co-operative groups; and emphasise reaching the correct answer as quickly as possible. The French approach to primary teaching, in other words, makes it difficult if not impossible for children to learn the skills of exploration, investigation and problem-solving which are seen as important to primary science in England. As in Germany, any science learning which goes on in primary schools is largely related to 'learning about' scientific matters.

Primary Science in Anglophone Africa

In contrast, the situation in many anglophone African countries is in principle almost the reverse of that in Europe, regarding their approach to teaching science to primary age children. After independence, many African countries realised the need for rapid economic growth and saw education, particularly science education, as a key element in this. Many countries committed huge proportions of their GNP to educational expansion, and in most cases science was made a compulsory examinable element of the primary curriculum from the outset, i.e. soon after independence. Aid agencies, particularly in Britain, North America and Scandinavia, were heavily involved in supporting curriculum development in primary science, and a process approach was at the heart of most developments. An example was the Science Education Programme for Africa (SEPA), based in Ghana, involving science educators from several West and East African countries, established in the mid-1970s. Many countries subsequently adopted the ideas and materials developed by such projects to suit their own country's needs, and generated process-based classroom materials for primary schools.

The development of primary science in Kenya is a good example of what happened throughout much of anglophone Africa. In the mid-1970s, the Kenya Institute of Education, supported by science educators of international standing, had redesigned the primary science curriculum along 'practical problem-solving' lines, and produced excellent Teachers Guides and up to 50 'Primary Science Units' for pupils, to support the changes in the classroom. The initiative was closely followed by changes in the primary leaving examination also geared to supporting the change; pioneering work was carried out to develop questions in science which would test the higher-order intellectual and process skills being fostered, work which was taking place before most primary teachers in Britain were even thinking about teaching science to their children. Later, commercial publishers in Kenya produced primary science texts to back up this approach (for example Berluti, 1980, 1989) which were of such high quality that I have consistently used them with learner teachers and on INSET courses here in Britain. All these changes were systematically and centrally introduced into the schools through the two-year initial teacher training programmes in the 18 primary teachers' colleges.

And yet not surprisingly, there is little sign that, fifteen years later, these changes have given Kenya any economic advantage over its neighbours, or even more importantly, little sign that the changes have had an impact on the post-basic secondary school curriculum. Clearly in economic terms, any effect of the changes has been totally masked by influences such as world commodity prices, political stability and tourist fashions. And not only are macro-economic and political factors suppressing such effects; cultural factors also play a crucial part. Russell, writing about Botswana, has shown how the process approach runs contrary to cultural norms of good learning (which involve, for example, not questioning authoritative sources), and is thus difficult to establish as a legitimate method of teaching (Russell, 1991). Jegede & Okebukola (1991) have also demonstrated how even the learning of a basic skill such as first-hand observation can be powerfully inhibited by traditional beliefs and taboos about the objects being observed.

The Impact of Change in Primary Science

Which compels me to ask the same questions of our own system, namely to what extent has the change in emphasis in primary science had an impact on science in secondary schools, and do cultural factors play the same subtle role in neutralising impact? We perhaps only need to look at the cultural significance of 'A' levels and their influence on curricula to realise that the same is indeed true. One anecdote may also add a disturbing illustration. In 1953, during my first week as a pupil in grammar school, my first science lesson was about 'the parts of a bunsen burner' which I faithfully copied from the board. 38 years later, after the expenditure of millions of pounds on science curriculum development and in-service training, when my son started at his comprehensive high school in the other corner of England, his first science lesson was also about 'the parts of the bunsen burner'. Whilst writing this article, therefore, I asked 139 first-year undergraduate learner-teachers about their first science lesson in secondary school. Around 90% reported the same bunsen burner lesson. All this despite the advent of GCSE and the National Curriculum, which perhaps illuminates dimly what might be going on in practice behind the facade of rapid and major change. This does not need to be taken as an indictment of secondary school science; but given the real changes in primary schools from little or no science to compulsory teaching of the subject, it may suggest that the way science is taught at the beginning of the secondary phase is in practice still impervious to 'bottom up' influences from primary schools.

So in placing equal stress on science as both a body of knowledge and a set of process skills which are of equal importance, we are not only trying to do something which has been shown to be extremely difficult if not impossible in practice (Claxton, 1991); we are also out of step with several European countries whose economies are arguably more successful than ours, and in step with many African countries who after two decades or more of promoting process science are as yet not reaping any observable benefits.

Primary Science and National Aspirations

It can be argued from this that the connection between a curriculum which emphasises early science learning of an investigative nature, and a nation's economic success, is at best not proven and at worst might be an expensive and over-ambitious endeavour. And it is made all the more worrying not only by the evidence that a process-oriented curriculum has always proved difficult to deliver, but also by the growing evidence from our own country that primary teachers lack confidence in teaching science (Wragg, Bennett & Carre, 1990); that INSET support does not seem to have long-term impact, after support is withdrawn (ASE, 1988); that national curriculum assessment of 7-year-olds analysed so far shows 'a lack of familiarity (amongst teachers) with the processes on which good practice in science is based' (CATS, 1991); and that the teaching and assessing of science in primary schools, as a consequence, is proving so onerous that the Secretary of State's requirements in this respect are undergoing gradual cutbacks.

There are many questions raised by this situation. On what aspects of their education system, if any, is the German success in science and technology based? Are we yet ready in this country to attempt to introduce 'process science' to young children, in terms of resources, teacher competence and assessment techniques? Or is it that behind the supportive consensus for investigative work with young children, there is an emperor with no clothes? Perhaps it could be argued of us, as a culture, that we have never really believed in the rhetoric of science education as the engine of economic improvement. Perhaps our track record of amateurism and the 'wacky inventor', are closer to what we as a society actually believe in as the way forward. Media stereotyping of scientists would certainly give strong support to such a view. Or perhaps it is all an elaborate way of blaming teachers and pupils for deep-seated inadequacies in other aspects of the nation's economic and social life. It could certainly be argued, right now, that our primary-age children might be better off learning a second European language.

In cultural terms, one peculiarly English phenomenon which might have implications for all these questions is to do with what one might call the 'permanently provisional' nature of most of our ideas about schooling. This was characterised in a TES editorial on the 'reading question', but it might equally have been said of several other curriculum subjects. The leader suggested that:

> The intense emotion associated with defining different views [of reading] derives from all sorts of deep conflicts: between schools and society; between common sense and expertise; between left and right; between different attitudes to education. (Should it be child-centred or subject-centred? Is it about process or outcomes?). We are never going to agree on a 'right way' (*Times Educational Supplement*, 28.9.90)

What struck me forcibly in European schools was that teachers had much less difficulty accepting that there might be a 'right way'. And colleagues working in other systems such as Hong Kong, for example, have reported the same situation amongst teachers, even during the current period of uncertainty. Perhaps in primary science our accepted consensus will turn out simply to be too ambitious. Patricia Broadfoot (1990) has hinted that our ideology of child-centredness may have to be sacrificed not because it is inappropriate as an ideology, but because of the current pressure to cover all the attainment targets, an example of a self-fulfilling political prophecy. Can a right way be right, if it becomes impracticable? Are we wise to aim high, and continue to fall short of our aspirations, whilst others set more modest targets for primary schooling and perhaps do better as a result? Or does our approach need celebrating, as an example of healthy pluralism, in contrast to European authoritarianism?

There will be no easy research answers to such questions, even though the partisans will continue to grind axes over them. Perhaps the more urgent issue for primary teachers is for research to establish whether the process skills which we set such store by are actually learned as a consequence of teaching, or acquired intuitively as children mature. If it turns out to be the case, as Preece (1984) has suggested, that many intuitive science ideas are 'triggered', innate concepts rather than learned, 'constructed' concepts, where does that leave the justifications for our emphasis?

References

Association for Science Education (1988) *Initiatives in Primary Science Education (IPSE): An Evaluation Report*. Hatfield ASE.

Berluti, A. (1980) *Beginning Science, Standards 4–8*. Nairobi, Kenya: Macmillan.

——(1989) *Science 6–14: An Activity and Resource Manual for Teachers*. Nairobi, Kenya: Macmillan.

Broadfoot, P. and Osborn, M. (1990) French Lessons. *Times Educational Supplement*. 9.11.1990, p. 12.

Claxton, G. (1991) *Educating the Enquiring Mind: The Challenge for School Science*. London: Harvester-Wheatsheaf.

Consortium for Assessment and Testing in Schools (CATS) (1991) *The Pilot Study of Standard Assessment Tasks for Key Stage 1*. SEAC.

Cornelsen-Velhagen and Klasing (1986) *CVK Sachbuch 2*. Berlin: CVK.

Department of Education and Science (1978) *Primary Education in England: A Survey by HM Inspectors of Schools*. London: HMSO.

——(1985) *Science 5–16: A Statement of Policy*. London: HMSO.

Jegede, O. J. and Okebukola, P. A. (1991) The relationship between African traditional cosmology and students' acquisition of a science process skill. *International Journal of Science Education* 13 (1), 37–47.

Krucker, F. -J. (1987) Heimatkunde or Sachunterricht? Concepts, queries and questions. In P. V. Dias (ed.) *Science, Society and Education: Basic Science at Elementary Education Level*. Frankfurt: Verlag fur Interkulturelle Kommunikation.

Kinder, K. and Harland, J. (1991) *The Impact of INSET: The Case of Primary Science*. NFER.

Lauterbach, R. and Frey, K. (1987) Primary science education in the Federal Republic of Germany. In P. V. Dias (ed.) *Science, Society and Education: Basic Science at Elementary Education Level*. Frankfurt: Verlag fur Interkulturelle Kommunikation.

Millar, N. and Driver, R. (1987) Beyond processes. *Studies in Science Education* 14, 33–62.

Peacock, A. (1989) What parents think about primary science. *Primary Science Review* 10, 20–21.

Preece, P. (1984) Intuitive science: Learned or triggered? *European Journal of Science Education* 6 (1), 7–10.

Russell, A. (1991) Primary science and the clash of cultures in a developing country. In A. Peacock (ed.) *Science in Primary Schools: The Multicultural Dimension*. London: Macmillan Education.

Times Educational Supplement (1990) Reading has to be first. *Times Educational Supplement* 28.9.90, p. 19.

Wragg, E. C., Bennett, S. N. and Carre, C. G. (1990) Primary teachers and the National Curriculum. *Research Papers in Education* 4 (3), 17–45.

PRIMARY PUPILS' CONCEPTIONS OF ENERGY: FACT OR FICTION?

Gill Nicholls

Christ Church College, North Holmes Road, Canterbury, CT1 1QU

Abstract Much has been written on children's alternative frameworks, prior conceptions, or understandings of scientific concepts. The realisation that these exist have made it clear that teachers and researchers need to take account of such notions when trying to design and develop teaching material. The need for innovative material is essential, if teachers are to be helped to implement the demands of the National Curriculum; where progression differentiation and continuity within the curriculum are necessary.

This paper suggests a possible underlying structure to the way primary pupils view energy. The data is explored and interpretations given to the results. The aim of the research was to see how pupils in the primary school conceived energy and energy related topics with the view of developing a teaching package that was to include computer software, and meet with the Attainment targets on energy in the science document of the National Curriculum.

Introduction

Nowhere has the need for research and development been more prominent than in the area of primary science and technology. The National Curriculum has brought to the fore front the need to consider the processes, skills, and conceptions young children have and need to cope with in Science and Technology. To be able to encompass such ideas into teaching/learning situations requires close scrutiny. The National Curriculum requires teachers to take account of *progression, differentiation, cohesion* and *continuity* within the curriculum. This, in turn, reflects what and how topics or knowledge areas are taught. If teachers are to make the most use of these notions they need to have a good understanding of the knowledge children have prior to teaching.

Extensive literature indicates that children come to their science lessons with prior conceptions that might differ from those taught in the classroom. (Bliss & Ogborn, 1985; Duit, 1981; Gilbert & Pope, 1986; Stead, 1980; Watts, 1983; Driver, 1983). These publications have left the researcher/teacher asking questions about how to incorporate these notions into effective teaching and learning strategies. A recurring theme is that learning is a process of constructing new knowledge on the basis of current knowledge. Running concurrently with this has been the advent of the computer into the classroom, adding to the need for research into how best to view its use and include it in the everyday teaching of science and other areas of the curriculum.

The present research project has focused on these issues when developing and integrating a piece of software for the teaching and learning of *energy* and *energy* related concepts with 9 to 11 year olds.

The study has so far shown that successful incorporation of software into the teaching of energy can be linked to children's prior conceptions and cognitive level. This paper considers the first of these two issues, that of prior conceptions. It will briefly indicate the theoretical framework from which the research was derived, and then examine how the pupil's prior conceptions were elicited and the conclusions drawn from the evidence. Finally it will give a brief description of how the evidence was used to develop a teaching package for this age range.

Theoretical Framework

Before embarking on such a research project a theoretical framework was required to guide the investigations into teaching and learning of energy using computer software. Four main areas were considered as important.

First was to understand how pupils viewed energy, on the assumption that the knowledge and ideas they held prior to teaching would shape their general understanding, appreciation and approach to the learning of ideas. Secondly a theory of learning was needed as a framework for the research, so as to have a basis for planning a teaching strategy, and describe consequent learning episodes. Thirdly to consider how useful a piece of software could be for the teaching strategy/learning process, with this in mind it was also necessary to consider the pupils cognitive skills with respect to those that might be incorporated into software and teaching learning tasks. As such the project encompasses:

(1) The role of the computer in teaching/learning structure.
(2) Teaching strategies and classroom learning.
(3) Children's ideas about energy.
(4) Cognitive development.

Eliciting Conceptions of Energy

Many of the recent publications on children's ideas of energy have attempted to explain the underlying causes and origins of these conceptions. One of the major descriptions of these is given by Watts (1983), who puts forward seven categories of energy. These can be considered as a set of metaphors to help understand children's ideas and explanations of energy associated with events. However these categories are difficult to use in terms of developing teaching material. What was required was to find if pupils aged 9 to 11 had an underlying structure to the way they conceived energy in order that a teaching aid could be developed to help the children progress with their ideas.

The position taken here for the research was that it was important to consider what the pupils already knew about energy, accepting that much of younger children's knowledge may be difficult to assess. It was therefore important to devise a research tool that could access the type of information required easily and simply from the children.

Developing the energy questionnaire

The questionnaire went through several stages of development. Initially a small group of primary and young secondary pupils (18) were asked a series of questions which required written responses. The type of questions asked included:

(1) What do you mean by energy?
(2) Where do you think energy comes from?
(3) How do we use energy?
(4) What sort of things do you think give energy?
(5) What sort of things do you think are energy?

The answers to the questions and the questions themselves were discussed with the pupils to find areas of difficulty and misunderstanding. The responses the pupils made were kept and used later to modify the questionnaire further. Two points became very clear at this juncture:

(1) Written responses would not elicit the necessary variety of aspects dealing with energy that was required,
(2) The pupils found it difficult to express their ideas in written form, they appeared to be more confident of their judgements rather than their reasons.

It was for these reasons a grid type questionnaire was developed requiring only yes/no answers about a variety of entities, asking for each a range of features related to energy. Nine aspects of energy were considered for 22 entities. All aspects chosen related to verbs most often used about energy when interviewed through the developmental stage of the questionnaire. Figure 1 relates to the 9 aspects of energy, and Figure 2 relates to the 22 entities.

The questionnaire was given to two classes of junior pupils ranging in age from 9+ to 11+, of mixed ability, prior to starting a six week Energy topic. The questionnaire was completed by all pupils at the same time.

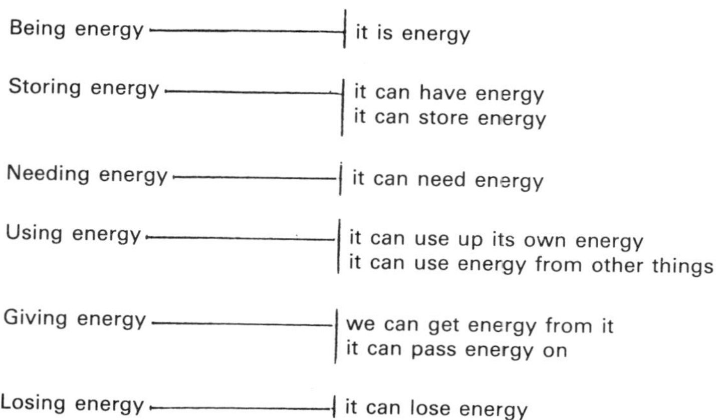

Figure 1 Aspects of energy chosen

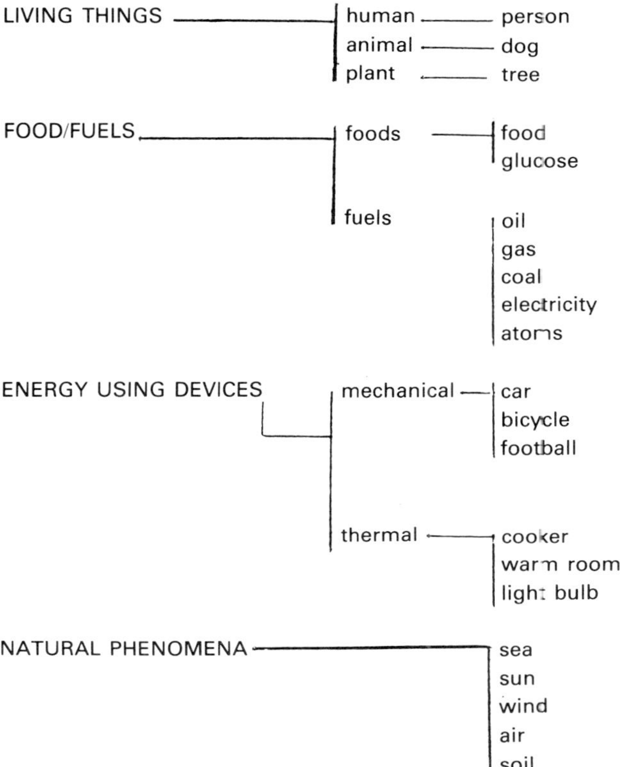

Figure 2 The entities chosen

Analysis of the Questionnaire

The main aim of the analysis was to see how and in what way the primary pupils perceived energy. The method of analysis chosen was Multidimensional Scaling (MDS). (For indepth discussion on this form of analysis refer to Nicholls, 1991).

This gives a map in two or more dimensions showing the distance between correlations between entities. For the younger pupils two dimensions explained 99% of the variance in the ordering of distances, with a low Kruskal stress of 0.024. Figure 3 shows the MDS map for the primary pupils.

It can be seen that there is a significant divide in the map horizontally, giving two distinct groups. However there is also a divide in the vertical dimension, in this way giving four groups. The interpretations given to the dimensions were identified by considering the percentage yes responses for all aspects of energy. These can be seen in Figure 4. This allowed for the possible characterisation of the four groups. (For ease of interpretation the entities have been re-ordered into four groups.)

From Figure 3 it can be seen that the main dimension (horizontal) has a clear interpretation. It includes the entities falling in the groups of '*living things*' and '*energy using devices*'. These fall to the right of the map. Comparing these with

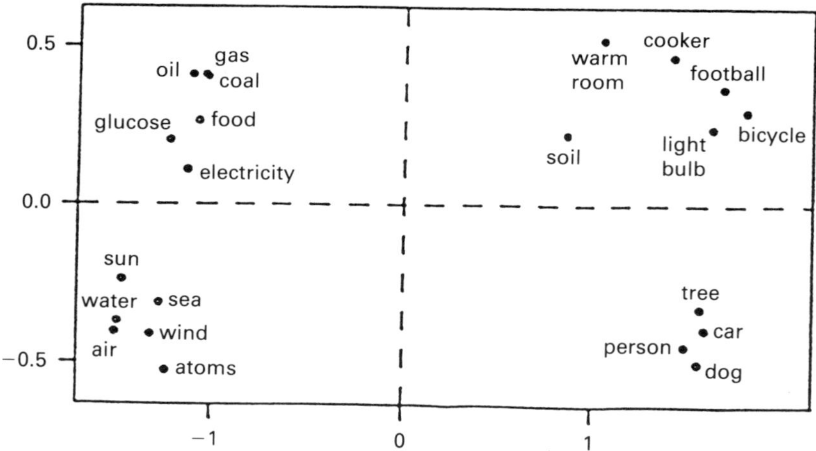

Figure 3 MDS map for primary pupils

Figure 4, it would seem that the pupils consider these entities as most often *'needing energy'* and *'using energy from other things'*, however they did not see the entities as *'giving energy, passing on energy, or are energy'*.

On the left of the map the groups of entities found represented *'foods and fuels'*, and *'natural phenomena'*. A complementary set of properties are represented in Figure 4. These two groups of entities are regarded by the pupils as being able to *'pass on energy'*, *'get energy from them'*, and *'are energy'*, and yet are rarely seen as *'needing energy'* or *'using energy from other things'*.

Using the above information the horizontal dimension of the MDS map was given the interpretation of CONSUMER/SOURCE whereby the *consumer* category includes 'living things' (person, dog, tree) plus car as well as energy using devices such as football, warm room and possibly soil. *Sources* were regarded as 'foods and fuels' (glucose, gas, oil, etc.) and 'natural phenomena' (wind, sun, sea).

The second dimension which is weaker appears to divide the entities in a more complex way. The interpretation given here is that the younger pupils could be seeing the entities in terms of 'act on own'/'used to act'.

This interpretation derives from the fact that 'living things' and 'natural phenomena' can be distinguished from 'foods and fuels' and 'energy using devices' by the aspect 'it can use up its own energy' (Figure 4).

What is being suggested here is that this dimension appears to be concerned with those entities that can or cannot use their own energy. Things that can 'act alone' are visibly seen to have energy, such as a person, dog, and car, they are all seen to move freely. Sources of energy such as sun, water and sea are all seen as visibly having energy.

Things that are 'used to act' could be interpreted as something needs to be done before energy is apparent. Each entity has to be acted upon before energy is visibly used, e.g. foods fuels, energy using devices.

The data was then subjected to a simple dot plan analysis which identified the aspects on which the entities in a given group were most frequently or rarely

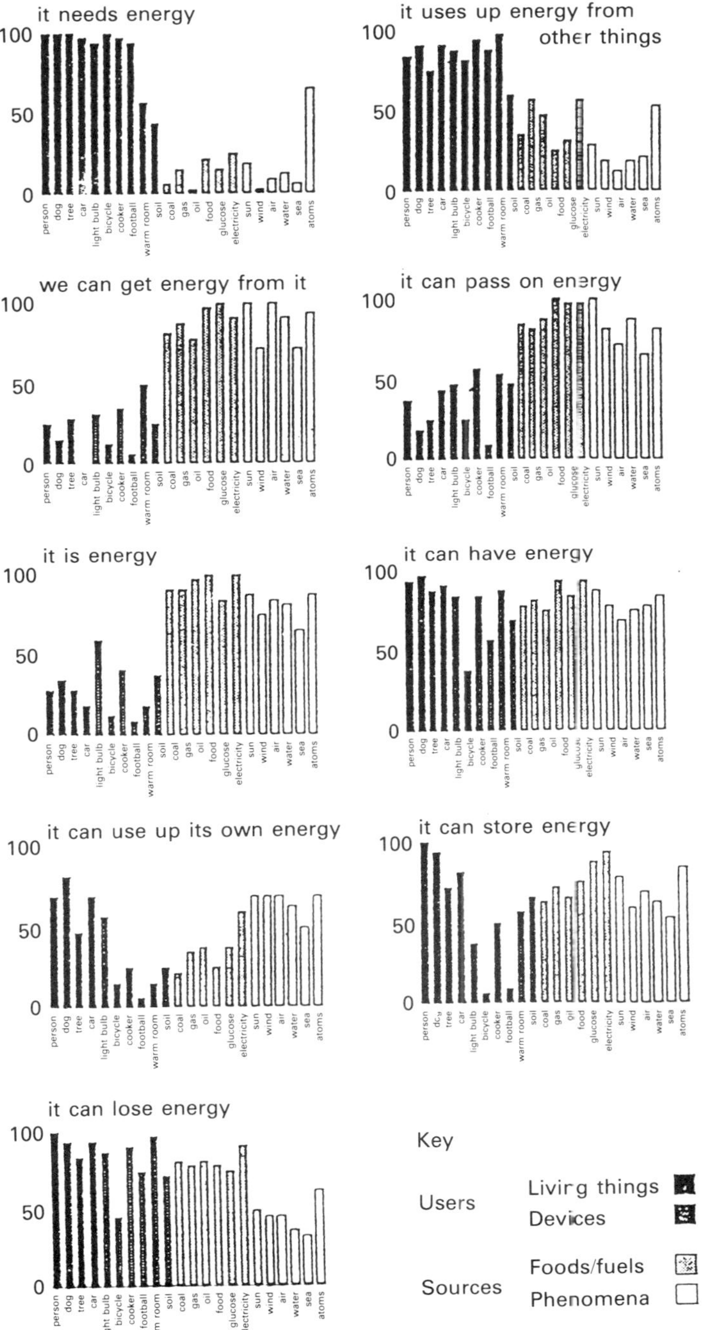

Figure 4 Percentage yes responses

Table 1 Dot plot analysis

SOURCES OFTEN SEEN AS	We get energy from it, it passes on energy.
CONSUMERS ARE RARELY SEEN AS	It is energy.
SOURCES ARE RARELY SEEN AS	It needs energy, uses up energy from other things, it loses energy.
CONSUMERS ARE OFTEN SEEN AS	Using energy from other things, needing energy.
ACT ALONE OFTEN SEEN AS	It uses its own energy, it can have energy, it can store energy.

Table 2 Summary of objects in four groups

Living things	Consumer	Act alone
Foods and fuels	Source	Use to act
Energy using devices	Consumer	Used to act
Natural phenomena	Source	Act on own

chosen. Table 1 gives an indication to how the notion of 'act on own' and 'used to act' were distinguished.

A summary of the nature of objects in the 4 groups can be seen in Table 2. Further analysis showed that three features could be identified. (Table 3). From Tables 1, 2 and 3, it can be seen that *energy using devices* are rarely seen as *storing energy*, and *natural phenomena* are rarely seen as *losing energy*.

It is possible to say that *users* which are also *used to act* do not store energy, while sources which *act alone* do *not lose energy*. This is also reflected in the second dimension where the entities are concerned with aspects of 'losing energy' and 'storing energy'. The findings are summarised in Figure 5.

Table 3 Identifiable features

Object Characteristics: (High = +) (Low = −)

Features	Giving Energy			Consumption/Loss			Possess Own Energy		
Groups	Get From	Pass On	Is	Need	Use Up	Lose	Use Own	Have	Store
Living things	−	−	−	+	+	+	+	+	+
Foods/fuels	+	+	+	−	−	−	−	−	−
Energy using Devices	−	−	−	+	+	+	−	−	−
Natural phenomena	+	+	+	−	−	−	+	+	+

NOT USE OWN

Foods and Fuels   do not store energy
                                                             energy using devices

SOURCE   USER

natural phenomena
do lose energy   living things

USE OWN

Figure 5 Summary of interpretations of two dimensions

Conclusions Drawn from Analysis

The analysis has shown that it is possible to describe underlying structures in the way young pupils conceive energy. A major feature has been in detecting a main structure to Primary pupils concepts of energy namely that they have a distinction between *sources* and *consumers* of energy.

Pupils appear to conceptualise sources as things which give energy, as well as being energy. These include foods, fuels, and visibly active phenomena such as wind, water and the sea. However it is important to identify the notion that pupils conceive a loss of energy with loss of activity, and distinguish between entities that 'act alone' and those that were 'used to act'. It is possible that this relationship stems from the younger pupils' view of animacy in association with energy.

Discussion

From the data and analysis put forward it would suggest that primary pupils have an underlying structure on the concept of energy. The question needing to be asked is how and in what way can this information be used?

The *fact* is that prior conception, alternative frameworks or whatever label we wish to give, these notions are very much with us. However the *fiction* would be to ignore the reality of such underlying structures amongst young pupils.

The project to date has developed a piece of software which is incorporated into a teaching package based on the data put forward in this paper. It emphasises starting from children's ideas and encourages the teacher to explore what these ideas are and how they can be used to negotiate the elements of energy as found in the National Curriculum, in terms of progression and differentiation, by giving cohesion and coherence to the theme of energy. The aim of the package is to progress pupils' conceptions of energy into a more scientific way of thinking.

The most important aspect is that it tries to promote learning within familiar experiences such as keeping warm, hence pupils can test and apply their knowledge and understanding easily.

Initial piloting indicates that it has potential for young pupils to conceptualise various elements of energy and energy related concepts within their own framework, yet encourages them into the scientific way of thinking about such ideas as heat flow, hot and cold and insulation. The results of this work are expected to be published early next year.

Acknowledgement

I would like to thank Professor J. Ogborn for all the help in initially analysing the data cited in this paper.

References

Bliss, J. and Ogborn, J. (1985) Children's choices of uses of energy. *European Journal of Science Education* 7, 195–203.

Duit, R. (1981) Students' notions about the energy concept—before and after physics instruction. In W. Jung, H. Pfundt and C. V. Rhoneck (eds) *Problems Concerning Students' Representations of Physics and Chemistry Knowledge* (pp 268–319). Proceedings of an International Workshop. Ludwigsburg: Pädagogische Hochschule.

Driver, R. (1983) *The Pupil as Scientist?* Milton Keynes: OU Press.

Gilbert, J. and Pope, M. (1986) *School Children Discussing Energy*. University of Surrey, Institution of Educational Development.

Nicholls, G. M. (1991) *Case Studies in the Use of Computer Software in the Teaching of Energy*. Unpublished PhD thesis. London: University of London.

Stead, B. (1980) *Energy*. Working Paper 17. University of Waitato: Learning in Science Project.

Watts, M. (1983) Some alternative views of energy. *Physics Education* 18, 213–17.

RESEARCH INTO ENGLISH PRIMARY SCHOOL TEACHERS' UNDERSTANDING OF THE CONCEPT *ENERGY*

Mike Summers

Westminster College, Oxford and Oxford University Department of Educational Studies, 15 Norham Gardens, Oxford OX2 6PY

and

Colin Kruger

Oxford University Department of Educational Studies, 15 Norham Gardens, Oxford OX2 6PY

Abstract Interviews-about-instances and interviews-about-events conducted with an opportunity sample of 20 primary school teachers in England were used to probe the teachers' knowledge and understanding of the concept *energy*. A number of statements made by the teachers during the interviews were then incorporated into a survey questionnaire which was given to an opportunity sample of 152 primary school teachers from three English Local Education Authorities. This enabled the prevalence of particular ideas about energy in a much larger sample of teachers to be established. Results of both the interview and survey phases of the research are reported, and discussed in light of a recent increased emphasis on the learning of science concepts in primary schools.

Introduction

The research reported here was undertaken in response to a concern about the ability of primary teachers to respond to the demands made by the introduction of a National Curriculum for science in England and Wales. A radical feature of this new curriculum is the shift from an emphasis on the processes of science towards a far greater concern for conceptual understanding at the primary level. In secondary schools the teaching of knowledge and understanding of the concepts of science has long been regarded as the task of subject specialists whereas in primary schools teachers are predominantly generalists and very few have received any formal education in science since the early days of their own secondary schooling (this is especially so in the case of physical science). The introduction of conceptual science into primary education raises two issues. The first of these is the range of levels to which younger children are able to deal with the more abstract ideas of science, if given the opportunity. There has been little research into the growth of primary children's powers to abstract and generalise within science (the UK Science Processes and Concept Exploration project, based at King's College London and the University of Liverpool, is a notable exception) and there is certainly plenty of scope for further research in this important area.

The second issue is the level of conceptual understanding of science possessed by primary teachers themselves. In the past few years the Primary School Teachers and Science (PSTS) Project has reported research into primary teachers' understanding of a number of scientific concepts including force and gravity (Kruger, Palacio & Summers, 1990a, b and c), materials (Kruger & Summers, 1989), and aspects of biological science (Lenton & McNeil, 1991a and b). This paper reports the findings for what is perhaps the most fundamental of all scientific concepts, namely *energy*.

Methodology

Like other PSTS research, the work on energy has involved an interview phase and a subsequent prevalence phase. The purpose of the interview phase is to explore in some depth the kinds of views of the targeted science concepts expressed by a fairly small sample of primary teachers (perhaps 20 or so people). In the prevalence phase, the intention is to find out how widely the views uncovered during the interviews are held in a large sample of teachers (typically 150 in PSTS research).

In the case of energy, in-depth interviews were conducted with an opportunity sample of 20 primary teachers from three English Local Education Authorities (LEAs). Teachers' understanding of the concept was probed using the interview-about-events (Osborne & Gilbert, 1980a) and interview-about-instances (Osborne & Gilbert, 1980b) techniques in which interviewees are shown either a real event (e.g. a jumping toy bug) or simple line drawings on cards (see below) and asked questions which focus on a particular aspect of the situations, in this case, the role of energy. Inferences are made from their responses about the nature of their understanding of the concept. The eight events and instances used to elicit teachers' views about energy are shown in Figure 1. In each case the main focus questions were 'Does the—have energy?' and 'Where does the energy come from?', followed by further questions framed according to the teachers' reply.

The instrument for the energy prevalence phase consisted of a questionnaire showing five of the drawings used in the interviews. Alongside each drawing was a selection of statements made by the interviewees about the situations represented. The interviewees' original words, modified slightly where necessary to ensure brevity and clarity, were used throughout. Scientifically correct statements about the drawings, using non-technical language, were intermingled with the interviewees' statements for each of the situations presented. These two types of statement were developed into the final version of the questionnaire (see Appendix) following trials with teachers during a pilot study.

The questionnaire, containing 26 statements in all, was completed by 152 practising primary school teachers from 85 schools in three LEAs during staff meetings, inservice courses and staff development days. All the teachers received the same standardised introduction to the questionnaire, which was administered in person by a member of the PSTS research team and collected immediately.

The rest of this paper is organised in four main parts. First, some of the more significant findings from the interviews concerning teachers' knowledge and understanding of energy will be described using quotations as supporting evidence

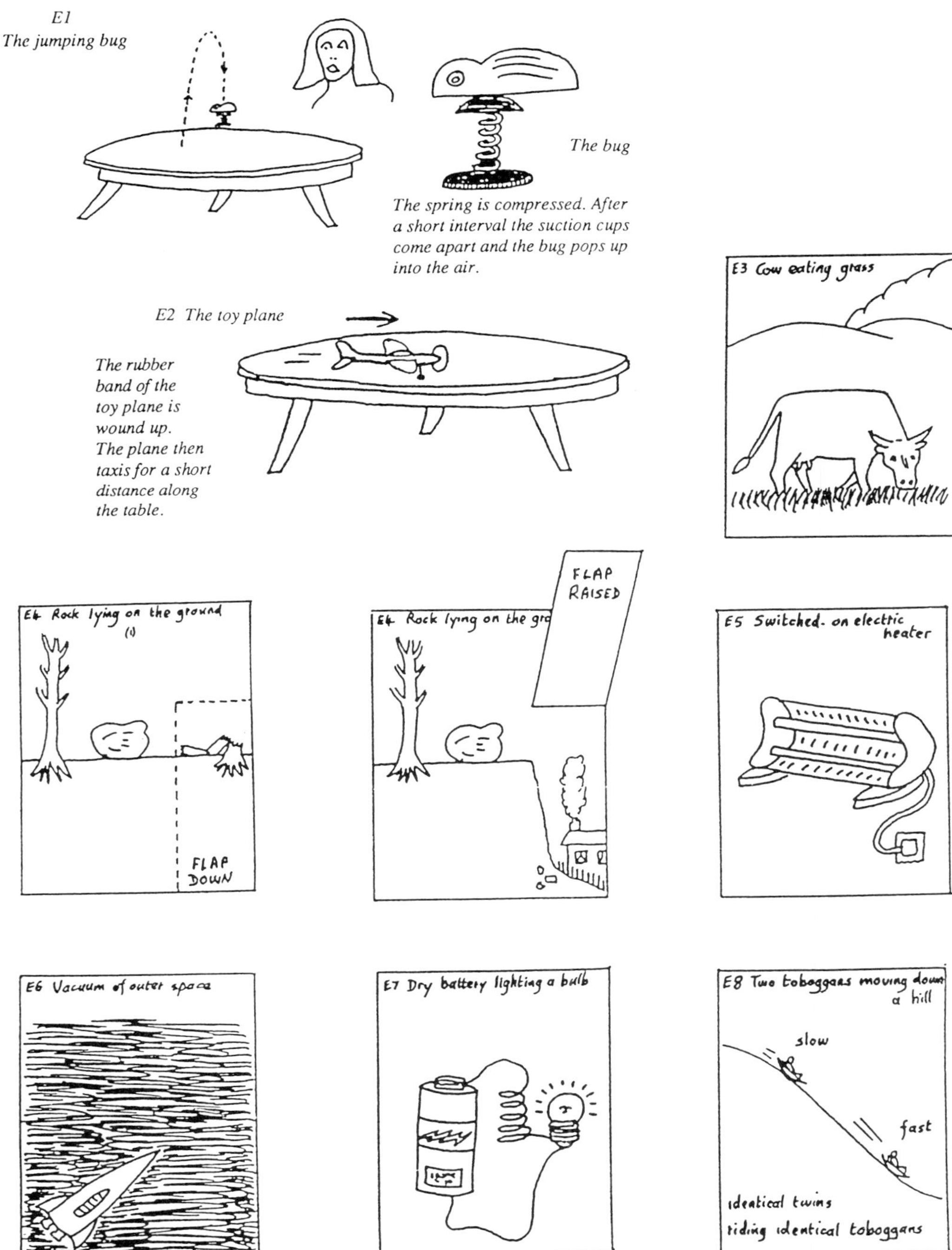

Figure 1 The eight events and instances used in the interviews.

for the interpretations offered. Second, some general characteristics of teachers' thinking about energy that became apparent from the interviews will be described, again using illustrative dialogue. Third, the questionnaire responses will be presented and briefly summarised. Finally, some of the implications for science in primary schools and the training of teachers will be discussed.

The Interviews: Knowledge and Understanding

In the following paragraphs figures in brackets within the text refer to the number of teachers holding a particular view (**I** denotes interviewer; **T** denotes teacher).

Energy and motion

Many of the teachers' notion of energy seemed to be mainly associated with movement and action; 13 of them, for example, attributed the cow's energy to its ability to move or do something. This resembled Watts' (1983) 'obvious' energy framework observed in children, where overt displays of activity are the sole means of identifying energy.

> **T**: . . . we need a definition of energy here . . . the ability to move, I presume, to move and to do things . . . (*interview 14, rock*)

A number of teachers did not regard stationary objects, such as the rock (13), or the bug on the table (16) or at the top of its flight (9) as possessing energy.

> **I**: . . . do you think it (rock at edge of cliff) has any energy?
> **T**: No, . . . it hasn't got any . . . way of moving itself. (*interview 19, rock*)

Three teachers were uncertain about the association between energy and motion.

> **T**: (the bug had energy) while it was going upwards . . . when it got to its maximum height the energy had been expended—that's the law of gravity.
> **I**: And after that it didn't have any energy? . . . (*teacher nods*) . . . OK. (*interview 7, bug*)

One teacher maintained that energy was present in the moving toys only until the spring and rubber band resumed their former shape and did not perceive a rock falling off a cliff as possessing energy. Two teachers identified the energy of movement but were not sure whether it increased with velocity.

> **T**: As it (the rock) gets faster? It gets more (energy) as it goes down? . . . No . . . I don't know. (*interview 13, rock*)

Energy and life

The teachers' idea of energy was often strongly associated with life; the cow was regarded by 9 teachers as possessing energy owing to its being alive.

> **I**: Does the cow have energy?
> **T**: Yes . . . It has energy because it has its own life force within it. (*interview 1, cow*)

Perhaps such views are displaying a relic of the 18th century theory of 'vitalism'.

> **T**: (The energy in your muscles comes) from your vital force really, from the food you eat and the, the living force. (*interview 19, bug*)

Several teachers did not consider that energy was a term applicable to objects if they were inanimate, whether they were moving, such as the toboggans (3), or stationary, such as the rock (5). Energy was also seen as human liveliness:

> **T**: . . . there's the sort of thing when you've got lots of energy, you can do a lot, you can move around, you can jump up and down . . . (*interview 5, toboggans*)

This is a view resembling children's 'human-centred' conceptualisation of energy identified by Watts (1983). Anthropomorphic explanations are characteristic of this perspective in children and these were also noted in some of the teachers' views, particularly in connection with the bug and plane instances.

> **T**: . . . it (the spring) wants to go back to its original shape . . . (*interview 18, bug*)

Energy and force

The 'active' notion of energy described above also contained an association with force. Several teachers were confused about the distinction between energy and force.

> **T**: . . . I'm probably mixing up energy with force and work . . . what is energy? (*interview 5, toboggans*)

Some used the two words synonymously (4) while others regarded friction (5) or gravity (2) as energy.

> **T**: There's a pull from gravity.
> **I**: Is that energy or is it a force or are the two the same?
> **T**: The two are the same. (*interview 4, toboggans*)

> **T**: It . . . has got an energy force of its own but it only has an energy force when something is done to it. (*interview 1, plane*)

Six of the teachers saw energy as a force present all the time within a substance.

> **T**: . . . I would say (energy is) a force contained in a particular thing, a mechanism or a substance . . . (*interview 17, bug*)

This force was hidden and was waiting to be used—the 'depository model' of energy where it is 'a causal agent, a source of activity' (Watts, 1983).

> **T**: . . . that was a hidden force within there (the bug's spring) . . . therefore I knew there was energy within there . . . (*interview 1, bug*)

Sources of energy

Three of the teachers gave metaphysical explanations of energy and saw it as proceeding ultimately from the Creator.

> **T**: . . . man cannot create energy, man may take energy from one source and

lighten another, but the creation of energy, that's God's work. (*interview 3, plane*)

Fifteen teachers named the sun as the original source of all energy, describing 'energy chains' involving fossil fuels or food.

> **T**: Well it's part of a chain because electricity is only part of what is produced in a power station which is produced from the coal what has then come from the sun . . . it all comes from the sun basically. (*interview 9, heater*)

This view is similar to the 'flow-transfer' energy framework which has been noted in children (Watts, 1983) in which energy is viewed like a fluid which is transported from one body to another. Few teachers mentioned photosynthesis (4) or the associated role of carbon dioxide (1) although some stated that carbon (1) or chlorophyll (3) were involved. So a number of these teachers may lack knowledge of the carbon cycle, which is fundamental to an understanding of global energy problems such as the 'greenhouse effect'.

Conservation of energy

Conservation was probed directly only in the case of the two toboggans; only a few teachers (4) explained the situation in terms of decreasing potential energy and increasing kinetic energy as the toboggans descended. The upper and lower systems were felt to have the same amount of energy by eight teachers altogether, while seven believed the bottom toboggan to have more energy because it was going faster. Other teachers could not answer conclusively or did not consider that energy was a term applicable to toboggans.

> **T**: Well I don't know that either of them has got energy. I don't think that toboggans have got energy. I don't think things falling down . . . friction or whatever . . . they were falling so they would create energy but I don't think they have got energy. (*interview 18, toboggans*)

Conservation also arose spontaneously in the course of discussion of some other instances and revealed some inconsistencies. For example, one teacher was both a 'conserver' and a 'non-conserver' of energy and felt that 'it depends what you're talking about'. Another could not resolve an inconsistency concerning energy 'loss' in outer space.

> **T**: . . . if you were out there in a space suit, you would have to be insulated because otherwise all your heat would go out into this thing (space) so it can soak it up—you would radiate heat out into it—exactly where that would go I don't know! (laughter)—there's a nothing to go out to so there's no energy in it . . . (*interview 10, astronaut*)

Several of the teachers (5) said that energy could be 'created' or implied in their responses that energy was not a conservable entity.

> **I**: Can you create energy—in the sense that there's no energy and suddenly it's there?
> **T**: I think so. (*interview 6, cow*)

Few teachers described details of energy transformations in the examples, only one referring to the transformation of energy into sound and a single other to dissipation into low-level energy.

I: (The energy you put into the plane was) used up? What do you mean by that? It just vanished, disappeared?
T: No it could have jiggled the air molecules about a little bit, it might have become heat . . . (*interview 4, plane*)

Potential energy

Most of the teachers lacked a clear concept of energy possessed by reason of a body's position. Some (6) used the term 'potential energy' but only one of these gave completely correct 'physics interpretations' of the situations. Others produced muddled statements such as:

T: the energy it (the bug at the top of its flight) had was less than gravity. (*interview 9, bug*)

Two believed that an object possessed gravitational potential energy right at the brink of a cliff (or table) but not at some distance in from the edge.

T: It seems to depend on the situation in which you place it . . . whether it's sitting on a table or whether it's on the edge.
I: So if it's on the table top it doesn't have any energy, but if it's on the edge, it does?
T: . . . that's what I would have said, yes. (*interview 6, rock*)

Only two teachers referred to the energy possessed by the bug and plane relative to the floor when on the table.

Other forms of potential energy were either not recognised by teachers or their understanding was not clear. Energy was identified by all of the teachers in the plane's rubber band and in the bug's spring (except for one teacher who did not perceive the compressed spring as possessing energy) and some called it 'stored energy' (5) or potential energy (2) but none named it as strain-energy or elastic potential energy. Chemical potential energy also was not named but was implied in responses such as 'energy from oxygen' (3), 'battery energy' (1), the 'stored energy' of food (3), fuel (4) and battery chemicals (7), or the energy of 'creation', thought by some teachers (6) to be inherent in all substances and extractable if they were processed in some way.

I: Does the rock have energy?
T: Mm . . . well it's created out of all the things it's created out of so I suppose it must have something latent inside it . . . there would be energy used in its creation . . . a piece of granite or . . . limestone . . . you . . . get the energy out of it . . . if it will burn . . . to a sufficiently great heat . . . (*interview 12, rock*)

A few teachers (5) had 'no idea' about the workings of a battery and, while some mentioned contents such as acid (4), lead (1), carbon (2), and 'alkaline' (2), only two talked about materials or chemicals reacting or interacting inside.

Conventional views about energy

Only one teacher gave totally correct interpretations of the role of energy in the situations presented in terms of the conventions of physics but two others did give definitions which contained a reference to work.

> T: . . . what is energy? . . . it's this power to work . . . I've done that to it
> (compressed it)—it reacted . . . (*interview 11, bug*)

> T: . . . in a way it's a potential to do work under your own steam. (*interview
> 5, toboggans*)

Some more general characteristics of the teachers' responses to the energy
instances and events now follow.

The Interviews: General Characteristics of the Teachers' Thinking about Energy

Linguistic difficulties

When talking about energy a number of the teachers used scientific jargon, for
example, in expressions such as 'the energy of gravity'. Such usage probably
reflects superficial knowledge rather than a particular view of energy. Some
teachers acknowledged the problem they had with the use of scientific language:

> T: I was dealing in words . . . which I may not have been using correctly.
> I: You had no real conception of what those words mean?
> T: No . . . I think it's . . . wrong to use words like that . . . (children) pick
> them up as correct language. (*interview 1, post-interview discussion*)

Two teachers pointed to a further linguistic difficulty—the breadth of meaning
they believed the word 'energy' possessed. Because of this one could not respond
to some questions with any conviction and the other proposed that, in some
circumstances, energy could be created.

> T: It's all down to how you define energy . . . there are so many ways to use
> it . . . it's a difficult one—I'm not always sure exactly how you are using the
> word energy . . . (*interview 17, heater, bulb*)

> I: Can you or coal or something just create its own energy, as it were?
> T: Yes I think it probably can because energy's a broad term, it must cover
> all sorts of things. (*interview 7, heater*)

Some teachers eschewed scientific language but others did use conventional scientific terms, such as kinetic energy (6), with varying degrees of confidence, due
possibly to their attempts (sometimes successful) to recall the symbolic world of
their school science lessons.

> T: . . . I can't remember the proper name for it (the energy in the elastic band
> of the toy plane). Going back to my scientific memory from secondary school,
> there are two sorts of energy but I can't remember the names . . . it's kinetic
> and what's the other name? One is stored in it and the other is energy it is
> using, I think.
>
> (*later*)
>
> . . . it has . . . potential so that when it (the rock) falls off the cliff it is then
> using up that energy. (*interview 9, plane, rock*)

Inconsistent and context-bound views

Many teachers' conceptual meanings for energy seemed to lack coherence and resulted in contradictory statements. For example, one teacher maintained that the jumping bug, falling rock and sliding toboggan had no energy during their movement, although the plane did have energy as it taxied along the table. Another believed that energy was connected with movement and also that movement could occur which did not involve energy.

> **I**: Is movement a form of energy?
> **T**: . . . you have to have energy there for movement to proceed, although some things can move without (it)—if they're thrown, they themselves haven't got energy . . . (*interview 5, bug*)

Such apparently inconsistent views expressed within the same interview have been observed in children's scientific thinking and seem to arise from the interviewee interpreting specific instances rather than invoking general laws (Driver, 1986).

Teachers' ideas as active schemes

Teachers' views were not static and often actively developed during the interview, with uncertainties being resolved. For example, two teachers who began by not associating energy with motion changed their view as the logic of their position was seen by them to be faulty during the course of the discussion.

> **T**: . . . I have to say (the rock's) still got no energy . . .
> **I**: it's on the edge of the cliff . . . just about to fall, does it have any energy then?
> **T**: . . . I'd still have to say no.
> **I**: . . . it started to fall down towards the house . . .
> **T**: I'd have to say no, but I'm not so sure now!
> **I**: When it got to the bottom, it was just about to hit the house, would it have any energy then?
> **T**: It's got to, hasn't it? (*interview 6, rock*)

These mature teachers did begin to change their views when 'forced' to talk about a situation and think it through. One who initially spoke of 'a potential to do something' and 'a potential to fall back to the floor' used 'potential energy' confidently later in the discussion. Another, beginning with the correct language but vague thinking, developed his ideas during the interview and showed clearer understanding at the end.

Restricted, perceptual and undifferentiated perspectives

The teachers who did not possess a developed idea of gravitational potential energy tended to restrict their thinking by focusing on properties of individual bodies rather than on their interactions within a system. So the rock or toboggan were thought to derive their energy from 'external surfaces' or 'friction' and their energy resulting from the context in which they were placed was not considered.

The tendency of some teachers to associate energy with life, action, movement and liveliness rather than with an inert state indicates that their thinking may be dominated by observable features in the problem situations presented to them. A characteristic of children's scientific thinking, noted by many workers during the past decade, is that it too is perceptually dominated, dealing with directly observable entities rather than with the application of abstract models to situations (Driver, 1986).

The intermingling of the concept of force with that of energy was evident in the combination of the two terms as 'energy-force' by teachers in those parts of the interview dealing with forces (see the section on energy and force above). Children's thinking has also been shown to contain undifferentiated notions in which separate scientific ideas or processes are merged into one.

Prevalence Phase

The prevalence questionnaire together with numerical responses of the 152 primary teachers in the sample is shown in the Appendix. Note that the statements are numbered from 73 to 96 because this particular questionnaire was one of three, each covering different science concepts. Statements 1 to 72 were on the questionnaires dealing with forces and changes in materials, and are not considered in this paper.

Jumping bug

Most of the teachers responded in a way consistent with the scientific views that the bug has energy when its spring is compressed and that it has energy when it is moving after the spring has uncoiled. However, two thirds of the sample manifested confusion between force and energy by agreeing that 'the spring's energy is a hidden force within it'. Almost all the teachers affirmed that the bug has energy when it is moving upwards but, surprisingly, nearly a third denied that it had energy when moving downwards. Over half the teachers believed that the bug possessed no energy at the top of its flight or were uncertain about this.

Although 60% of the sample affirmed that the bug's energy was its ability to do work, the research team believes on reflection that this statement 'stood out' too obviously as 'scientific' and that little store should be placed in the large number of 'correct' responses.

The final question about the bug probed knowledge of the notion of gravitational potential energy. Only 20% of the teachers agreed with the colloquially expressed description of the bug's potential energy with respect to the floor—more than half denied that the bug has such energy.

Toboggans

Just under a half of the teachers thought that the energy of the toboggans comes only from their movement, a further indication of the widespread lack of the concept of gravitational potential energy. However, an appreciation of energy conservation is implied in the responses of the 86 teachers who felt that both toboggans have the same energy if friction with the snow is ignored. This is consistent with a fairly large number of teachers denying that the faster, lower toboggan must have more energy than the top one (statement 82). Most of the teachers (82%) rejected the idea that energy is associated only with living things, although 25 (16%) thought that this was true or were unsure. The final toboggan question also indicated that a small number of teachers were unsure about energy and living things, with 12 agreeing with, or expressing uncertainty about, the statement 'the only energy in the picture is the liveliness of the twins'.

Electric heater

Two thirds of the sample affirmed implicitly that energy is conserved by denying that it did not exist before it was generated at the power station. However, this still means that more than 30% thought the energy did not exist previously or were unsure about this. The transformation of energy from one form into another was almost unanimously affirmed (statement 85), but very few teachers agreed with the scientific view that energy is an abstract idea. In fact more than two thirds denied that 'energy has no physical existence'.

Rock on the cliff

Again, a large proportion (66%) of the respondents did not seem to have (or were unsure of) the notion of gravitational potential energy (see statement 87). Also, and in agreement with responses to similar questions above, relatively few teachers (21%) implied by their responses to statement 88 that they felt (or were uncertain) that only living things could have energy.

The confusion between force and energy remarked on above in the case of the jumping bug was further evidenced by the 131 teachers who agreed that 'the falling rock has force'.

A lack of appreciation of gravitational potential energy emerged yet again in statement 90, where 45% of the teachers denied that a stationary rock at the brink of a cliff has energy and a further 22% were uncertain about this. However, once the rock had 'passed the point of balance' and started to fall, a large majority (78%) *did* affirm that it had energy. Nearly half of the teachers felt that the energy of the rock increases as it falls, while 29 were uncertain and 49 thought this was false. The implication is that only these latter 49 teachers have (or may have) some appreciation of the principle of conservation of energy.

Grazing cow

There was general unanimity about the grass as a source of energy for the cow and about the soil not being the sole source of energy for the grass. Almost all the teachers denied that a cow's energy is related only to its activity, but 69% of the respondents showed the vitalistic belief that 'the cow has life-force energy which is within all living things'.

Consistency of knowledge and understanding

The consistency with which teachers held a particular belief or lacked knowledge of a particular idea was investigated by analysing responses across contexts. So, for example, in the case of gravitational potential energy 81 of the 152 teachers indicated by their *false* response to statement 87 about the rock that they did not recognise this idea. Of these 81 teachers, 55 responded *false* to statement 78 about the bug indicating a consistent failure to recognise gravitational potential energy across these two contexts. Similarly, of the 68 teachers who apparently denied potential energy by responding *false* to statement 90 about the rock, 56 failed to recognise the same idea in statement 81 (the same context) and 46 in statement 78 (the bug). These figures are clearly indicative of a consistent absence of this particular idea. Still further consistency checks were carried out for gravitational potential energy (using other contexts), and for apparent recognition of conservation of energy and an association of energy with living things. This kind of analysis showed that, while many teachers did respond inconsistently, it is reasonable to infer that a substantial number were exhibiting consistency in their knowledge and understanding of the ideas probed. Further details of this analysis can be found in Summers (1991c).

Summary

Some of the more prevalent conceptual difficulties evident from the questionnaire responses were, then, as follows:

- a confusion between force and energy;
- uncertainty about the energy associated with an object as it moves upwards, reaches the top of its flight and then falls (very widespread confusion about the *forces* acting in this situation has been found in other PSTS research—see, for example, Kruger, Palacio & Summers, 1990c);
- an absence among many teachers of any notion of gravitational potential energy and an association of energy with motion rather than position;
- substantial numbers of responses contradicting the principle of conservation of energy;
- a widespread view of energy as a substantive entity rather than an abstract idea;
- widespread endorsement of a vitalistic view of energy as 'life-force' which is within all living things.

An association of energy solely with living things was not widespread, but up to about 20% of the teachers seemed to hold this belief or were uncertain, depending on the situation.

These results confirm that the lack of knowledge and understanding of the concept of energy which emerged during the small sample interview phase of the research are widespread in the larger community of primary school teachers. The final section of this paper discusses some of the implications of these findings for primary school science.

Discussion

The notion of energy is both one of the most important and one of the most difficult concepts in the whole of science. One reason for the difficulty is the abstract nature of the idea:

> Energy does not exist. It is nothing. It is a 'way to look at' something happen. (Guidoni, 1985)

> Energy is a most abstract idea, because it is a mathematical principle . . . a numerical quantity that does not change . . . is not a description of a mechanism, or anything concrete . . . (Feynman, 1963)

Another reason is that the biologist's (or chemist's) view of energy is unlikely to be identical with that of a physicist. This is not because the concept is different in the different fields of enquiry, but because physicists, chemists and biologists are likely to emphasise different aspects of the same underlying idea. It is particularly difficult to deal with energy in a purely qualitative way since the whole importance and meaning of the concept derives from its mathematical, quantitative formulation (see the quote from Feynman above). A further reason for the difficulties presented by the concept of energy is the broad socially acquired meaning of the word. The difficulties children have in dissociating the symbolic scientific meaning of energy from the wealth of meanings the term has acquired from 'Life-World' usage are well known (Solomon, 1983a). The results of the PSTS Project's interview study with primary teachers show that these problems are not confined to children.

How necessary is it to teach children about energy? Undoubtedly the concept of energy is of major importance throughout science. Also some understanding of ideas about energy is important for adult life and essential for understanding and making informed decisions about national energy policy. So the case for introducing children to the notion of energy as soon as it is possible for them to cope with such an abstraction is a strong one.

But the crucial question is when should the teaching of energy commence? The little research undertaken into children's readiness to conceptualise energy suggests that this cannot begin until the age of 13 to 14 years (Solomon, 1986). Other studies support the view that abstraction of scientific ideas is a process that commences in middle adolescence (Clough *et al.*, 1987). Furthermore, it has been stated that to teach any concept, particularly that of energy, before the proper foundations have been laid is positively harmful and that it is absurd to believe that energy can be understood by young children who are known to be incapable of abstract thinking (Warren, 1986). If these views are accepted it would seem that the task of the primary school is to provide a wealth of experiences and practical exploration for children upon which the later idea of energy can be built.

However the National Curriculum In England and Wales requires the development of primary children's knowledge and understanding rather than merely being exposed to a range of experiences. The National Curriculum Statements of Attainment for the primary age-range (DES, 1991), for example, include:

> 'be able to describe how a toy with a simple mechanism, which moves and stores energy, works'   (5 to 7 years)
> 'understand how energy is used and transferred in models and toys'   (5 to 7 years)
> 'know that there is a range of fuels that can be used to provide energy'   (5 to 7 years)
> 'understand that the Sun is the major energy source for the Earth'   (7 to 11 years)
> 'understand that energy is conserved'   (7 to 11 years)

It is naive to suppose that children in the primary school enjoying and discussing the range of experiences provided by teachers will not be developing an understanding of the word 'energy' which their teachers will so frequently be using, or will not be fitting the word into a 'schema' of some kind. It is equally naive to believe that primary teachers can effectively plan these experiences for children or utilise them in their teaching unless they have an appropriate understanding of energy themselves. The teachers' perspectives and difficulties illustrated through PSTS research are not surprising in view of their lack of scientific background. However, there is clearly a need for teachers' ideas to be moved away from these perspectives and towards the scientific view if they are to lead children towards a coherent view of energy and assess their level of understanding. These problems have to be addressed during initial and in-service training if science education in primary schools is to move forwards.

However, there is a proud tradition in primary schools of 'teachers learning with the children', and the reason why this approach is unlikely to be successful in the case of primary science in the future needs to be spelt out. The key point here is that scientific knowledge is not the result of common sense reasoning and science concepts cannot be induced directly from experience. Rather, present day science has arisen from imaginative theoretical constructions of some of our greatest scientific minds over the centuries. The conceptual journey from the young child's precursor notion of energy as 'personal liveliness' to the more complex scientific conception appropriate for a well-informed adult citizen is an arduous one and will require a guide who knows the route from its beginning and has some idea of the eventual destination. This is a powerful argument for primary teachers to develop scientific understanding of energy and to know of the pitfalls which lie in the clutter of 'messy, contradictory and obstinately persistent' Life-World views (Solomon, 1983b) that children possess. Only then will teachers be able to plan effectively these early experiences for children and pose the questions to help develop pupils' understanding. PSTS research would seem to indicate that most primary teachers accept as axiomatic the proposition that concepts cannot be taught, by whatever method, unless the teacher in the first place possesses an adequate level of understanding of those concepts (see, in particular, Kruger, Palacio & Summers, 1990d).

The entry of children into the Symbolic World (Solomon, 1983a) of science, whenever that may be, is more likely to be hastened and facilitated by an informed progression of experience and discussion rather than through a haphazard set of experiences from which it is hoped some kind of idea will emerge. Clearly, given the research findings reported above, much needs to be done in the training of teachers if the aims of the National Curriculum, with regard to energy, are to be achieved satisfactorily in our primary schools.

But what kind of teacher education is necessary to achieve this goal? The failure of science education in the past to provide the majority of pupils with a real understanding of science concepts has been demonstrated by a large body of research in the last decade and is now widely recognised. Mere transfer of conventional secondary methodologies to primary teacher education is unlikely to be a useful strategy, especially given the very different context of adult learning and the nature of the 'audience' (themselves experts in teaching and learning, if not in science). At a time of national innovation in primary science, the development and evaluation of new approaches are important concerns. Over the past two years the PSTS project has developed research based primary teacher science education materials rooted in a constructivist pedagogy and making extensive use of analogy in the qualitative learning of science concepts (Summers, 1991a and b; 1992). A longitudinal study of the long term impact of the strategies employed in these materials is now underway in the areas of *force* and *energy*. This will provide what is probably the first formal evidence in the United Kingdom of the extent to which primary teachers can develop an appropriate understanding of science concepts and thereby help meet the challenges now facing the teaching and learning of science in primary schools.

Acknowledgement

We would like to thank David Palacio and Joan Solomon for helpful comments on the energy questionnaire.

References

Clough, E. E., Driver, R. and Wood-Robinson, C. (1987) How do children's scientific ideas change over time? *The School Science Review* 69 (247), 255–67.

DES (1991) *Science for Ages 5 to 16 (1991)*. London: Department of Education and Science and the Welsh Office.

Driver, R. (1986) Restructuring the physics curriculum: Some implications of studies on learning for curriculum development. In Koichi Shimoda and Tae Ryu (eds) *International Conference on Trends In Physics Education—Proceedings* (pp. 9–40). Sophia University, Tokyo, 24–29th August 1986. Physics Education Society of Japan/International Commission on Physics Education.

Feynman, R. (1963) *The Feynman Lectures on Physics: Volume 1* (eds: R. Feynman, R. B. Leighton and M. Sands). Reading, USA: Addison-Wesley.

Guidoni, P. (1985) Energy matters. In *Proceedings of a Conference on Teaching Energy within the Secondary Curriculum*. United Kingdom: University of Leeds.

Kruger, C., Palacio, D. and Summers, M. (1990a) A survey of primary school teachers' conceptions of force and motion. *Educational Research* 32 (2), 83–95.

——(1990b) Adding forces—a target for primary science INSET. *British Journal of In-service Education* 16 (1), 45–52.

——(1990c) An investigation of some English primary school teachers' understanding of the concepts force and gravity. *British Educational Research Journal* 16 (4), 383–97.

——(1990d) INSET for primary science in the National Curriculum: Are the real needs of teachers' perceived? *Journal of Education for Teaching* 16 (2), 133–46.

Kruger, C. and Summers, M. (1989) An investigation of some primary teachers' understanding of changes in materials. *School Science Review* 71, No. 255, 17–27.

Lenton, G. and McNeil, J. (1991a) *Prevalence Phase—Results from the Biology Questionnaire. Working Paper No. 13. PSTS Project*. Oxford, United Kingdom: Oxford University Department of Educational Studies/Westminster College.

——(1991b) *Selected Findings from the Biology Interviews and Questionnaire. Working Paper No. 14. PSTS Project*. Oxford, United Kingdom: Oxford University Department of Educational Studies/Westminster College.

Osborne, R. and Gilbert, J. (1980a) A method for the investigation of concept understanding in science. *European Journal of Science Education* 2, 311–21.

——(1980b) A technique for exploring students' views of the world. *Physics Education* 15, 376–9.

Solomon, J. (1983a) Learning about energy: How pupils think in two domains. *European Journal of Science Education* 5, 49–59.

——(1983b) Messy, contradictory and obstinately persistent: A study of children's out-of-school ideas about energy. *School Science Review* 65, 225–33.

——(1985) Teaching the conservation of energy. *Physics Education* 20, 165–70.

——(1986) When should we start teaching physics? *Physics Education* 21, 152–4.

Summers, M. (1991a) *The Forces INSET Materials: Structure and Evaluation. Working Paper No. 12. PSTS Project*. Oxford, United Kingdom: Oxford University Department of Educational Studies/Westminster College.

——(1991b) *The Energy INSET Materials: Structure and Evaluation. Working Paper No. 15. PSTS Project*. Oxford, United Kingdom: Oxford University Department of Educational Studies/Westminster College.

——(1991c) *Inference of Consistency in the Prevalence Phase: Microcomputer Analysis. PSTS Project working document*. Oxford, United Kingdom: Oxford University Department of Educational Studies/Westminster College (in preparation).

——(1992) Improving primary school teachers' understanding of science concepts—theory into practice. *International Journal of Science Education* 14 (1), 25–40.

Warren, J. W. (1986) At what stage should energy be taught? *Physics Education* 21, 154–6.

Watts, D. (1983) Some alternative views of energy. *Physics Education* 18, 213–17.

Appendix: The Energy Questionnaire

The statements below were alongside line drawings of the situations under consideration in the version of the questionnaire completed by teachers. These line drawings were the same as those shown in Figure 1 of the main article. The frequency of the teachers' responses to each statement is shown in bracket (numbers of teachers who made no response to a statement are not included).

The picture shows a toy 'jumping bug'. The person compresses the spring so that the suction cups stick together and places the bug on the table. After a short time the suction cups come apart, releasing the spring, and the bug pops up into the air and then falls back onto the table.

71. When the bug's spring is compressed, but before it 'pops' up, the toy does not have energy.   **true (26) false (117) don't understand (0) not sure (9)**

72. When it's moving the bug does not have energy after the spring has uncoiled.
true (36) false (108) don't understand (1) not sure (7)

73. The spring's energy is a hidden force within it.
true (110) false (19) don't understand (6) not sure (17)

74. The bug has energy when it's moving upwards.
true (129) false (9) don't understand (1) not sure (11)

75. At the top of its flight, when the bug is moving neither up nor down, it has no energy.   **true (46) false (66) don't understand (2) not sure (37)**

76. During its flight the bug's energy is its ability to do work.
true (92) false (20) don't understand (18) not sure (19)

77. The bug has no energy when it's moving downwards.
true (45) false (92) don't understand (0) not sure (15)

78. When it is at rest on the table the bug has energy because it would fall to the floor if the table were removed.
true (31) false (84) don't understand (10) not sure (27)

Each toboggan started from the same place and in the same way—the twin didn't push it but just sat on it and it started to roll. The top twin has just started going down the hill but the other one further down started earlier and is going much faster.

79. The only energy that the toboggans have comes from their movement.
true (68) false (66) don't understand (3) not sure (15)

80. Both toboggans have the same energy, if you ignore their friction with the snow.   **true (86) false (40) don't understand (7) not sure (19)**

81. Toboggans have no energy because only living things can have energy.
true (11) false (124) don't understand (3) not sure (14)

82. If it's going faster, the lower toboggan must have more energy than the top one.   **true (31) false (93) don't understand (3) not sure (23)**

83. The only energy in the picture is the 'liveliness' of the twins.
    true (7) false (130) don't understand (5) not sure (7)

The picture shows an electric fire plugged into the wall. It's switched on and the bars are glowing.

84. The energy from the power station which supplies this heater did not exist before it was generated at the station.
    true (35) false (103) don't understand (2) not sure (11)

85. The heater is changing electrical energy into heat and light.
    true (143) false (3) don't understand (0) not sure (6)

86. Unlike force, which you can feel, energy has no physical existence since it is merely an abstract idea.
    true (14) false (96) don't understand (11) not sure (30)

This picture shows a rock lying near the edge of a cliff with a house at the bottom. The land erodes away until the rock is right at the edge of the cliff. Further erosion causes it to fall down the cliff.

87. Since it can't do anything the stationary rock initially doesn't have energy.
    true (81) false (48) don't understand (0) not sure (22)

88. The rock has no energy because it's not a living thing.
    true (14) false (115) don't understand (2) not sure (18)

89. The falling rock has force.
    true (131) false (7) don't understand (4) not sure (10)

90. Following erosion, the rock has energy when it's stationary right at the brink of the cliff.   **true (43) false (68) don't understand (5) not sure (34)**

91. The rock has energy when it's passed the point of balance and starts to fall.
    true (118) false (17) don't understand (1) not sure (14)

92. As the rock falls its energy increases.
    true (69) false (49) don't understand (1) not sure (29)

The picture shows a cow grazing in a field.

93. The cow gets some of its energy from the grass it eats.
    true (147) false (5) don't understand (0) not sure (0)

94. Grass gets its energy only from the soil.
    true (17) false (129) don't understand (1) not sure (4)

95. The cow only has energy if it is doing something.
    true (6) false (142) don't understand (1) not sure (3)

96. The cow has 'life-force' energy which is within all livings things.
    true (105) false (5) don't understand (22) not sure (19)

TACKLING CONTRADICTIONS IN TEACHERS' UNDERSTANDING OF GRAVITY AND AIR RESISTANCE

Robin G. Smith and Graham Peacock

Centre for Science Education, Sheffield City Polytechnic, 36 Collegiate Crescent, Sheffield S10 2BP, UK

Abstract Concerns over teachers' knowledge of science, particularly their understanding of forces, are reviewed. The article reports a study which explored the understanding of gravity and air resistance among a group of primary teachers on an in-service course. Two main misconceptions were identified. The first, that heavy objects fall more quickly than light ones, seemed to arise from seeing the effect of air resistance on the descent of light objects. The second, that gravity pulls all objects equally, seemed to arise from the observation that all objects which are heavy for their size fall at approximately the same rate. The idea that the forces from air resistance and gravity are balanced when terminal velocity is reached was also resisted by the teachers who felt that an object must be stationary when there is no resultant force. A sequence of teaching which tackled these misconceptions and the teachers' difficulties with balanced forces was evaluated. The discussion considers ways in which such work may best be done with primary teachers to increase the knowledge base of the profession.

Introduction

In the past few years there has been a greater emphasis on the teaching of scientific knowledge and understanding to children in the primary school. The swing away from an over-emphasis on process was heralded in Harlen (1978), Black (1980) and Kerr & Engel (1980). This became established in policy through *Science 5–16: A Statement of Policy* (DES, 1985). The introduction of the National Curriculum (DES, 1989) speeded up the change to a primary science curriculum where process and knowledge are seen to be of equal importance. However there has been growing concern that primary teachers themselves may not have the knowledge to teach this. Lack of such knowledge was cited by HMI as a constraint on primary science before these developments (DES, 1978). Anecdotal evidence and statistics showing primary teachers' qualifications (e.g. Keys, 1984) supported that. Teachers themselves have frequently volunteered a concern over their lack of scientific knowledge, especially in aspects of physical science, and their anxiety peaked as National Curriculum programmes were introduced (Carré & Carter, 1990). Recent studies suggest that in fact primary teachers are ill-equipped with scientific knowledge that could underpin their teaching (e.g. Kruger & Summers, 1990; Kruger, Palacio & Summers, 1990). More precisely, such studies have identified areas where most teachers hold ideas other than the standard scientific ones and make little use of scientific models to explain the world. One such area is the understanding of forces, including gravity.

The study reported here examined the impact of a short programme of teaching on the understanding of gravity and air resistance among a group of primary teachers. They were all science coordinators attending a DES-funded 20 day course run by a higher education institution for a local education authority (referred to here as Dodd LEA).

The concern over primary teachers' knowledge led the DES to earmark money for courses of 20 days' duration designed to upgrade their scientific understanding:

> The main aim of the courses should be to develop teachers' confidence and ability in the subject knowledge and understanding which will enable them to teach the programme of study for key stages 1 and 2 (DES, 1990)

The introduction of the courses has required the development of appropriate teaching approaches and materials. The emphasis in policy and in curriculum development has been on scientific understanding. The research into primary teachers' knowledge has focused on common scientific misconceptions. It has largely ignored the wider field of research on subject knowledge in teaching.

Subject knowledge itself needs to be analysed. Subject knowledge was dubbed the 'missing paradigm' in educational research by Shulman (1986). Subsequently researchers have begun to explore the knowledge base teachers have and what additional knowledge they might need.

The more sophisticated research programmes have conceptualised teachers' professional knowledge as involving other components such as general pedagogical knowledge, curriculum knowledge and understanding of learning in interaction with subject knowledge. The growth and use of that repertoire of knowledge has been analysed in some longitudinal programmes (notably Shulman's project at Stanford). Ideally we need similar studies of primary teachers' knowledge for teaching science. However there is also a need for detailed examination of features such as their own understanding of particular aspects of science which they have to teach.

What science understanding do primary teachers have?

Recent studies suggest that primary teachers' understanding of scientific concepts may be tenuous even in areas where they feel confident, such as the natural sciences. In the course of a survey about children's ideas about the concept of 'animal' Osborne & Freyberg (1985) found that only 86% of experienced primary teachers said that spiders or worms were animals.

Teachers express much greater concern over their ability to teach aspects of physical science. In a survey of 901 primary teachers by the Leverhulme Primary Project 74% felt competent to teach pupils how to look after living things but only 44% felt they had the knowledge to teach them how to construct circuits and only 45% felt their understanding of forces equipped them to teach it to pupils (Wragg & Bennett, 1989).

In the context of this present study one of the teacher participants asked his colleagues about their confidence in teaching particular areas of science. The results he found are summarised in Table 1. Clearly in this school the teachers perceived forces as being the area they felt least confident to teach.

Table 1 The areas of concern about science for teaches in one primary school

Content area	Not confident			Very confident		Total 1–3
	1	2	3	4	5	
Life and living			18	45	36	18
Earth and environment			27	64	19	27
Forces	9	27	36	18	9	72
Energy		27	27	36	9	54
Materials			9	73	18	36
Electricity		9	27	45	18	36
Investigating		11	33	55		44

All figures are percentages

Teachers' understanding of force and the implications for practice

The difficulties associated with teaching force and the need to examine these to improve practice were pointed out some time ago by Warren (1979:45):

> The idea of force is of very great importance in elementary science. It is obviously very widely misunderstood not only by students but also by highly qualified, mature adults. It is hard to assess how intrinsically difficult the idea actually is since there is a confusion of approaches to teaching it. Paradoxically, if we were to recognise that it is difficult, and were to teach accordingly, the subject would become more easy.

Exploring conceptual understanding in science

In the past two decades there has been an increasing body of research evidence which suggests that learners approach scientific work with theories of their own (e.g. West & Fensham, 1974; Driver, 1983). These theories are generally internally consistent and resistant to change. An example of such a set of beliefs is quoted in Finegold & Gorsky (1991:97) who suggest that the most commonly held views about force and motion can be summarised as:

> If a body is not moving then there is no force acting on it.
> If a body is moving then there is a force acting in the direction of movement.
> Constant motion requires constant force.

There have been several large scale projects which have sought to use a variety of research techniques to probe the understanding of key scientific ideas. Most have dealt with children: the Learning in Science Project based in New Zealand with Osborne *et al.* (LISP); the Children's Learning in Science Project at Leeds with Driver *et al.* (CLISP); Science Processes and Concept Exploration at Liverpool with Russel *et al.* (SPACE). More recently there have been a number of projects which have sought to probe the understanding of adults. These include the Primary School Teachers and Science project based in Oxford with Kruger *et al.* (PSTS). One of the main concerns which has prompted this last inquiry has been with the level of science knowledge of primary school teachers who are faced with teaching the full range of science in the National Curriculum.

Most of these projects have used variations on the techniques described by

Osborne & Freyberg (1985:8) as 'interviews about instances' which are highly specific and concern the children's use of words. They also employed the looser 'interviews about events' which involved the use of line drawings to prompt discussion and response. Later projects such as PSTS use similar techniques whilst SPACE uses a wider range which includes concept mapping, labelled diagrams, diaries, group log books and completing a picture.

Kruger, Palacio & Summers (1990) investigated the ideas about force held by 159 primary teachers. In their sample none of the teachers held a true Newtonian view of force. Most of their respondents had the view that force was something which was intrinsic to the moving object and that motion stopped because the object ran out of impetus. This idea has considerable implications for teaching about gravity.

Dealing specifically with gravity and weight Kruger, Palacio & Summers (1990) found that about one third of the respondents thought that gravity was 'a force produced by the atmosphere'. None of their respondents mentioned the conventional connection between gravity and weight. They noted it as a source of confusion and 'in nearly half of the interviews gravity was said to be a separate force acting in addition to weight' (1990:398).

Kruger & Summers (1990:93) detail some of the implications of their research into teachers' concepts of force. They suggest that teachers need to be taught about force as an external pull or push, the relationship between weight and gravity, and that an object at rest or moving at a constant speed does not have any force acting on it. They suggest useful strategies for helping teachers to come to terms with these ideas including the importance of making the knowledge of direct relevance to the teacher's own classroom work.

Kruger, Palacio & Summers (1990) in their work on force and gravity noted that all the teachers in their sample actively wanted to improve their own understanding. Kruger *et al.* also suggest that the idea that it is necessary for teachers of young children to stay just one step ahead of the children is dangerous. If teachers are only one step ahead then they will have no idea about the eventual aim of their teaching and the planning of a progressive scheme for teaching will be impossible. Kruger *et al.* wonder to what level teachers' understanding should be taken. They are rightly sceptical about the idea that GCSE level is about the correct level. There is no shortage of evidence that pupils, and teachers, who have studied science to GCSE or beyond may still cling to alternative conceptions which could hinder their teaching.

The earliest discussions of how primary teachers might cope with the science National Curriculum often revolved around questions about what level of the National Curriculum they needed to achieve themselves to teach to the lower level which their pupils might attain. Whether the argument is couched in terms of GCSE or National Curriculum levels it cannot be assumed that teachers simply require a slightly higher level of knowledge across the subject. Different elements of their teaching will need different degrees of understanding of underlying concepts. The mathematical demands of some aspects at a higher level may be quite unhelpful to the teacher. In some topics a width of knowledge and awareness of many different representations and explanations are likely to be crucial for teaching rather than a sophisticated scientific language. In others a deeper grasp of quite difficult concepts, albeit only in qualitative terms, may be needed by teachers who have to choose appropriate teaching approaches and answer pupil's questions. Forces may well fall into this category.

What children need to know about gravity

The details of the science National Curriculum have gone through a series of revisions. This may have been driven largely by concerns about assessment but the saga also illustrated a variety of views about scientific knowledge for pupils. At the time of this study the original orders (DES, 1989) were in use and there was a separate attainment target concerned with Forces. The statements of attainment which were the basis for assessment included two at level 4, relevant to primary teachers:

> Pupils should understand that things fall because of a force of attraction towards the centre of the Earth.
> Pupils should be able to recognise that weight is a force, and know that it is measured in newtons.

The programme of study which is the legal basis for planning also explicitly stated that children in key stage two should:

> Explore the forces which involve movement, friction and falling under gravitational force.

The following ideas are a summary of what we feel is implied by those extracts in the context of other references to forces, for instance to balanced forces, in the National Curriculum:

By the end of key stage two children will understand that:

- The force of gravity pulls things towards the centre of the Earth.
- Gravity is a pull which exerts a force on a mass. This force is measured in newtons.
- Where an equal force is operating in the opposite direction to gravity then there is no resultant force on the object. This is the case whether or not the object is in motion.
- When objects fall they accelerate.
- Objects which are light for their size e.g. feathers and parachutes stop accelerating because of air resistance. They then fall at a constant rate.
- All objects which are heavy for their size will accelerate at an equal rate. All objects which are heavy for their size will reach the ground simultaneously. (This will be the case when the heights are relatively small.)

Subsequent revisions of the Science National Curriculum (NCC, 1991; DES, 1991) have reduced the statements of attainment and made some modifications to the programmes of study. These might be interpreted as slightly reducing the expectations of what pupils should learn about gravity and forces—for instance by removing reference to weight as a force measured in newtons. However there is now specific mention of gravity in the key stage one programme. We would argue that the underlying ideas about gravity and other forces which children are expected to learn are similar. Children will in any case continue to drop things and experience gravity. We suspect also that teachers will continue to use activities where objects such as seeds, paper or parachutes are dropped. They therefore need to understand something of the forces which are involved.

What teachers need to know about gravity

If children should tentatively understand the above ideas about gravity then teachers should feel equipped to deal with the contradictions which are inherent in the learner's view of gravity. We suggest that these contradictions are:

- A heavy object like a big ball of clay falls at the same rate as a light object such as a paperclip. However, heavy objects are attracted to the earth with a greater force than lighter objects. At first sight these two observations are in conflict. The contradiction is resolved when it is realised that the heavy object needs a larger force to make it move. The crucial extra idea here is that of inertia.
- An object which is very light for its size such as a feather falls at a constant rate because the air resistance is equal to the force of gravity. The contradiction most people experience is that if two forces are equal then there should be no motion. The resolution lies in understanding that an object moving at a steady speed in a constant direction has *balanced* forces acting on it.

Prior to the teaching none of the primary teachers in this study were able to resolve these contradictions.

Aims of the inquiry

The present study was designed to explore several related questions:

- What did teachers actually know about the force of gravity before they started to study it on the course?
- What were the contradictions which teachers recognised in their own understanding about gravity?
- Was it helpful to explore these contradictions?

Methodology

The sample

All the teachers mentioned here have been given pseudonyms. The sample of fourteen teachers from Dodd LEA represents an opportunity sample. They are not necessarily representative of primary teachers as a whole. Most of the teachers were from primary schools but three were from middle schools which take children up to thirteen years old. Six of the teachers are from first schools which take children up to eight or nine years old. One of the teachers was trained as a secondary school teacher of biology after taking a degree in that subject. Another teacher took science as the main subject of his teaching certificate. All were responsible for coordinating science in their schools.

Methods

Following a pilot phase in which interviews and questionnaires were trialled with other teachers there were three stages to the study. The data collection methods used in the course of the inquiry were:

In stage 1:

* Structured interviews and interviews about events.
* Drawings made by teachers.
* Questionnaire.
* Interviews to expand on the questionnaire.

In stage 2:

* The teaching session.
* Observation of teachers during the teaching session.

In stage 3:

* Teachers' written and verbal responses to the taught session.

Stage 1: The sample group's existing understanding of gravity

Interview

Two teachers were chosen at random to be interviewed. They were drawn from teachers in the study group. The tutor had worked with them for five days previous to the work on gravity and had built up a good working relationship. The questionnaire which was filled in by all the teachers on the course was used as part of the interview.

It was decided at an early stage to tape record all interviews. Permission was always sought from the respondent. This method was supplemented by the use of rough notes. The interviews were located where possible in quiet rooms where a degree of privacy was possible.

Respondent A is a music graduate whose father is a professor of science. He did a PGCE immediately after leaving university. He has worked in a first school for five years and has had responsibility for science for two years. The other members of the course seemed to think of him as being knowledgeable about science. He took and passed physics O level.

Respondent B has taught for over fifteen years. She admits to being terrified by science and said she was 'dismayed' when she was asked to be the science co-ordinator at her infant school. It was clear from her reaction to the course that she gained very rapidly in confidence as she saw that the others whom she thought were 'so confident' are really at much the same level as she is.

Questionnaire

The questions used in the questionnaire and the results are discussed at the start of the next section. Since the tutor was with the students after they had filled in their questionnaire he took the opportunity to discuss the responses. The students were then asked to write comments in the light of this discussion without altering their original answer.

Stage 2: Teaching

The teaching session

The model for planning this work was derived from the 'generative learning' model suggested by Cosgrove & Osborne (1985:82). This provided a clear structure for exploring existing ideas and challenging them in a supportive manner (see Table 2).

The teaching material was developed to be of use not only to the researcher but more widely as a teaching resource for subsequent use by colleagues. The practical work in the teaching session had the following sequence:

(1) Dropping light things.
   Exploration of the part that air resistance plays in the way light objects fall.
(2) Dropping paper.
   Experiments with changing the amount of air resistance but not the weight.
(3) Dropping small objects which are heavy for their size.
   Changing the weight and air resistance of paperclips.
(4) Dropping feathers.
   Experimenting with terminal velocity.
(5) Dropping heavy objects.
   Experimenting with fairly heavy objects and their rate of fall.
(6) Drawing the threads together.
   Drawing out the contradictions in understanding of force and the rate of fall.
(7) Exploring proportions.
   Experimenting with paper and reducing the weight, mass and air resistance in exactly the same proportions.

Table 2 Generative teaching model applied to this inquiry

Preliminary phase	Concentrated work by the course tutor on his own conceptual understanding of gravity prior to the teaching session. Elicitation of the learners' general ideas of gravity using a concept map and a brief summary of their ideas about gravity.
Focus phase	The course tutor provided many activities which relied on the use of everyday objects. All experiences related back to things the learners had heard about, seen or felt. Learners were asked to discuss their ideas with a small group.
Challenge phase	All learners asked to contribute to an open discussion session where all ideas were debated openly. The course tutor described and demonstrated the scientist's view. He also suggested new lines of activity and 'thought experiments' to illustrate particular points. Learners compared their ideas about gravity with the scientist's view.
Application phase	Course tutor distributed the line drawing prompts from Osborne (1980). Learners covered up the accepted view and debated the answer in pairs. Delight reigned when most answered all the highly abstract ideas correctly using Newtonian mechanics. (Also some despair at ever being able to convince colleagues that Newton was right about the way things move!) A whole class debate over problems about terminal velocity. Whole class worked at this and came up with many problems and solutions related to real objects like feathers.

Commentary on each step in the teaching sequence

(1) The objective underlying the first activity involving dropping very light objects was to show that not all light objects dropped in the same way. There are chances for the learner to say what they felt caused the difference in the way the objects fell. The activity is a chance to warn teachers not to do this activity to illustrate gravity to their classes since the air resistance 'swamps' the effects of gravity.

(2) The second activity in the sequence allows learners to manipulate the variable of surface area whilst keeping the weight the same. It was important to draw learners' attention to this factor.

(3) The third activity which involved dropping paperclips with pieces attached gives an opportunity to see that when weight was added the objects fell more slowly. This challenges the learners' previously held assumption that more weight means a faster drop.

(4) The activity with feathers explores the contradiction that forces are always acting in the direction of motion. This is the clearest example the tutor could think of where an object was moving but balanced forces were in operation.

(5) This simple activity gave learners the chance to try something that many of them had never done. The application here is a 'thought experiment'.

(6) This activity emphasises that large masses are pulled by a greater force than small masses. The tutor felt that this direct approach was necessary as it was most unlikely that learners would discover this for themselves.

(7) This dealt with the idea that mass and gravitational attraction are in proportion. Interestingly this idea was talked about by teacher A in the taped interview before he undertook the activities.

Stage 3: Teachers' written and verbal responses to the taught session

The teaching session with the 14 Dodd LEA teachers lasted two hours. It was very lively with course members working through the gravity activities and materials in pairs asking frequent questions of the tutor who made notes and tape recorded as many of the conversations as possible.

Immediately after the session and as part of the learning experience the course members were asked to write one side about their responses to the teaching session. They were invited to say what they found most helpful and what things they found most difficult to understand. They were also asked to say what they still felt they needed to tackle.

Analysis

Questionnaires

Fourteen questionnaires were distributed and returned before the teaching session. The teachers all said they found them enjoyable to complete.

The first series of questions which asked about the forces on a ball thrown a little way in the air posed many problems. It was adapted from Osborne, Schollum & Hill (1981:3). The question is reproduced as Figure 1.

The first series of three questions was designed to test the teachers' general understanding of the forces operating on moving objects. The response generally

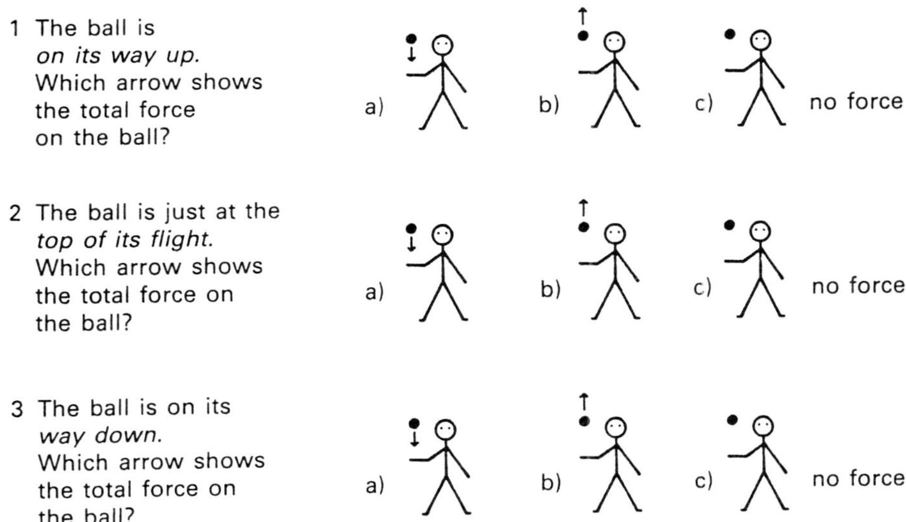

Figure 1 Questions about the forces on a ball in flight

accepted by scientists is a-a-a. Scientists refer to Newton's laws of motion and gravity from which it is clear that there is no force *in* the ball itself. Once it has left the person's hand the only force acting all the time is gravity. Only one of the teachers in this study answered that gravity was acting downwards in all three illustrations. Most of the respondents said that there was a force in the direction of movement (responses bca).

When commenting on her own responses one of the respondents thought that 'speed (upwards) was a force'. Another two respondents, again commenting on Q1, thought that 'Your force was stronger than gravity'. The picture with the ball at the top of its flight provoked the most difficulty and comment. In later discussion four of the respondents correctly identified the direction of the force. Another four of the respondents reported thinking that since the ball was stationary when it reached the top of its flight then the forces were cancelled out.

The next section of the questionnaire sought to analyse the course members' ideas about the force of gravity acting on objects of differing mass (Figure 2).

Only three out of fourteen respondents in this study gave the accepted view that the objects were pulled differently by the force of gravity. The other 10 respondents gave reasons like:

'I decided to say they were pulled the same because they fell at the same speed.'
'They travel at the same speed so have the same force.'

One of the respondents said:

'I have a contradiction of thoughts . . . all things fall at the same speed . . . but commonsense says that heavy weights fall faster.'

The third question on the questionnaire asked whether people thought that a large mass would fall faster than a small mass.

These objects have *just* been detached from their supports.

Draw lines to show the force on each object.

Use long lines to show big forces.

Short lines to show small forces.

Figure 2 Gravitational force on objects

Four respondents said that it was the heaviest first, two said it was the lightest first and seven gave the accepted view that they would reach the ground at the same time.

The reason one respondent gave for saying that the lightest would fall quickest was 'it has less air resistance so falls faster'. One of the respondents who replied that the heaviest would fall faster has a degree in science. She said that she could remember $F = ma$. A larger mass would have a larger force so it would fall faster.

Results from interviews with two teachers

Teacher A, who had studied science up to 'O' level, outlined his ideas about gravity in terms of what he had picked up at school. He quickly brought in the idea that air resistance slowed things. He contrasted the weight of a ball bearing with that of a feather. He said that the ball bearing would fall faster than a plastic straw and attributed this to the weight and to the idea that there was a 'speed, a speed which it can't fall if you like, it can't fall at supersonic speed there's a threshold beyond which it can't fall'. Teacher A suggested that heavier objects would be pulled more by gravity but that the force was the same 'per cubic metre'.

When asked about a feather falling in air Teacher A said that 'gravity would pull it down' and there would be a force from the air pushing up. When asked about the size of the forces Teacher A said that the force down must be greater than the force up 'or it would be hovering'.

Teacher B had not studied science past her third year at secondary school. She felt herself to be very uncertain about science and did not like to show her ignorance in front of others. Teacher B said that she had been told that it was not the weight which caused some things to fall faster than others but she then said 'surely if something's heavier it will fall faster'. Teacher B said that part of her theory came from watching weighing scales where the heavy side comes down. She was certain that larger masses were pulled down harder than the small masses. She went on to say that this confirmed what she said in the questionnaire about the heavy one hitting the ground before the light one. When pressed about the idea that all things fall at the same rate Teacher B said that she 'just ignored it, I shied away from it because I know I don't cope with scientific principles very well'.

Classroom observations and discussions

The main contradictions seemed to emerge when the students found out by experiment that all objects which are fairly heavy for their size fell at the same rate. They found it difficult to make sense of this in the light of the idea that large masses were attracted with more force than small masses. They became agitated and tried at first to resist the second idea but accepted it after considering the reading on spring balances. All the taped conversations and notes support the idea that it is important to develop an open and supportive atmosphere for learners to expose their ideas to scrutiny. Teacher A was adamant:

> Now that I've talked it round yes. Yesterday I was talking rubbish (referring to interview). Once you've got all the ideas out you can clear all the blocks and see the contradictions. All the talk has helped.

Later Teacher A went on to say:

> We've got lots of contradictory concepts we have to weave through. I think talking about contradictions is an enlightening approach. You've got to get people to admit they were wrong. It's OK. It's the best approach.

Teacher D felt that his experience in struggling to learn these ideas would affect his own teaching. 'You teach things better when you've really worked (on it). You skate over the things you find easy'. Teacher A independently supported this: 'You're not aware of the problems if you're a genius. It's like music. Geniuses can't teach' (Teacher A is a musician.) Teacher M also said that he felt that the session and the approach would help him to admit to his pupils that at times he didn't know some things.

There follows a sample of other responses to the approach adopted:

> I like the activities for each little bit. You can follow through that easily.
> I liked the questionnaire. Having done that there was a gap in my knowledge. It was helpful.
> I didn't know what I didn't know. Yes that's a gap in my knowledge I didn't know what I was doing there was a gap in my knowledge.
> If a child asks 'why' then you need the background knowledge.

Discussion

This short piece of action and inquiry has been reported to highlight some issues we need to tackle in assisting primary teachers to teach science. The accumulation of such accounts and the transfer of insights among those working with teachers can build understanding of how best to deal with those issues. The development of teaching materials and approaches may be the most enduring effect of courses such as that cited here. It is highly unlikely that even those teachers who are able to attend such courses will feel that 20 days supplies all their needs for teaching scientific understanding. In any case we are still at an early stage in teaching scientific ideas to all primary pupils. There is much to be done in primary science.

Although no generalised claims can be made on the basis of this limited study the findings were consistent with others which have found a number of non-Newtonian views among teachers. Even in such a small sample different degrees of understanding were found. Elicitation of individual ideas is necessary if we aim to help teachers review them as a basis for building more standard scientific frameworks.

Specific topics such as gravity may present particular problems. They may be inherently difficult and include apparent contradictions. The scientific ideas may be hard to illustrate with first hand experience. To explain them adequately we may have to introduce other ideas or use a quantitative approach; this can alienate teachers who view themselves as unscientific. Other topics may be much more accessible. We should therefore be wary of making any general statement about the level of scientific understanding required to give teachers the confidence and ability to teach through primary school. It may need to be higher in some carefully selected key ideas. Inherently difficult material will need to be taught in a carefully structured way. We need to research with teachers the areas they find harder to learn and to teach. Sometimes these will be in the context of familiar activities which only reveal difficulties when analysed. Investigations involving dropping objects such as seeds and parachutes fall into this category.

However there is a danger of disabling teachers by exposing contradictions and their misconceptions. This could simply reinforce their awareness of gaps in their knowledge and their view that physics is hard. Whatever support is provided should be designed to give them some success and confidence and enhance their ability to apply knowledge in teaching situations. We are concerned that research and inservice education with teachers may be emphasising a deficit model by focusing on unscientific concepts they use. There are some less threatening routes into this by exploring children's ideas and their assessment for the National Curriculum. It is also always refreshing to point to evidence that the alternative ideas of students beyond the primary stage can be resistant to the teaching of standard concepts by scientifically qualified teachers!

So while many teachers, including those on the course in this study, welcome challenges to their own scientific thinking others may be less happy to explore their own uncertainties. Many of those will also prefer to work from materials which are more directly applicable to their classroom teaching. Whatever the approach all teachers will need a positive climate to elicit and talk through their ideas. The tutors facilitating this need a bank of suitable teaching strategies as well as science knowledge. They need to ensure that the understanding which teachers do bring is identified and valued. They require empathy and sensitivity to probe contradictions and encourage growing understanding. They have to decide when it is appropriate to provide information and explanations. There should be adequate back up material to consolidate the standard scientific view. We must not leave teachers confused or resentful.

Although many science educators would now subscribe to a constructivist approach there are still questions about how this can be interpreted in practice with teachers. Where there have been thorough explorations with teachers of their subject knowledge and the related pedagogical knowledge these have taken time (e.g. Smith & Neale, 1987). The extent of the scientific ideas to be taught also indicates the need for some direct teaching and private study. However we wonder how far teachers can really use distance materials in isolation to develop their knowledge base sufficiently. It is also not clear that teachers can easily pass on to colleagues in school the understanding they have acquired themselves through working hard with others under the guidance of a tutor. There is no quick fix for the problem of a shortage of scientific knowledge among primary teachers.

However there are many positive steps. The experience of current inservice courses and research will yield analogies and representations that are productive to use with primary teachers. It will evaluate activities and questions which are assumed to be helpful. Follow up work is needed to pursue the integration of teachers' subject knowledge into their existing professional knowledge. We should not forget that teachers' expertise includes the use of subject knowledge, but only as one element in their practical knowledge. We need to keep inquiring about the purposes that knowledge might serve.

McDiarmid, Ball & Anderson (1988) cite evidence that teachers' subject matter understanding is critical for their capacity to pose questions, select tasks, evaluate pupils' understanding and make curricular decisions. They suggest that beginning teachers must learn to be flexible in choosing appropriate representations of their subject. This, they argue, requires a much deeper understanding of the subject matter than is needed simply to tell pupils what they ought to know. Beginners also have much to learn about pupils, teaching, schools and the curriculum as well. Our experienced primary teachers already use a wealth of such knowledge. However many are now grappling with the deeper knowledge of science that all teachers will need to give pupils their entitlement. To incorporate such knowledge into their repertoire they may need to test it in virtual and real practice, i.e. through planning and teaching (Schon, 1987; Smith, 1988).

References

Black, P. (1980) Why hasn't it worked? *Times Educational Supplement* 3.10.80.

Carré, C. and Carter, D. (1990) Primary teachers' self-perceptions concerning the implementation of the National Curriculum for Science in the U.K. *International Journal of Science Education* 4, 327–41.

Cosgrove, M. and Osborne, R. (1985) Lesson frameworks for changing children's ideas. In R. Osborne and P. Freyberg. *Children's Learning in Science.* London: Heinemann.

DES (1978) *Primary Education in England: A Survey by HM Inspectors of Schools.* London: HMSO.

——(1985) *Science 5–16: A Statement of Policy.* London: HMSO.

——(1989) *Science in the National Curriculum.* London: HMSO.

——(1990) *Revised criteria for designated courses in maths and sciences for primary teachers.* 6 August 1990.

——(1991) *Science for ages 5–16 (proposals).* DES.

Driver, R. (1983) *The Pupil as Scientist.* Oxford: OUP.

Finegold, M. and Gorsky, P. (1991) Students' concepts of force as applied to related physical systems. *International Journal of Science Education* 13 (1), 97–113.

Harlen, W. (1978) Does content matter in primary science? *School Science Review* 59 614–25.

Kerr, J. and Engel, E. (1980) Can Science be taught in primary schools. In Richards C. and Holford, D. (1985) *The Teaching of Primary Sciences: Policy and Practice.* London: Falmer Press.

Keys, W. (1984) *Aspects of Science Education in English Schools.* NFER Nelson.

Kruger, C. and Summers, M. (1990) A survey of primary teachers' conception of force and motion. *Educational Research* 32 (2), 83–95.

Kruger, C., Palacio, D. and Summers, M. (1990) An investigation of some English primary school teachers' understanding of the concepts force and gravity. *British Educational Research Journal* 16 (4), 383–97.

McDiarmid, G. W., Ball, D. L. and Anderson, G. W. (1988) Why staying one chapter ahead doesn't really work: Subject specific pedagogy. In M. C. Reynolds *Knowledge Base for the Beginning Teacher*. Oxford: Pergamon Press.

NCC (1991) *Science for Ages 5 to 16 (Proposals)*. London: NCC.

Osborne, R. (1980) *Learning in Science Project: Force*. Waikato: University of Waikato.

Osborne, R., Schollum, B. and Hill, G. (1981) *Force, Friction and Gravity: Notes for Teachers*. Waikato: University of Waikato.

Osborne, R. and Freyaberg, P. (1985) *Children's Learning in Science*. London: Heinemann.

Schon, D. A. (1987) *Educating the Reflective Practitioner*. San Francisco & London: Jossey Bass.

Shulman, L. (1986) Those who understand: Knowledge growth in teaching. *Educational Researcher* 15, 4–14.

Smith, R. G. (1988) *Thinking and Practice in Primary Science Classrooms: A Case Study*. Unpublished PhD, University of Leeds.

Smith, D. C. and Neale, D. C. (1987) *The Construction of Expertise in Primary Science*. Paper to AERA Annual Conference April 1987.

Warren, J. W. (1979) *Understanding Force*. London: John Murray.

West, L. and Fensham, P. (1974) Prior knowledge and the learning of science. *Studies in Science Education* 1, 61–82.

Wragg, E. C. and Bennett, S. N. (1989) Primary teachers and the national curriculum. *Research Papers in Education* 4, 3.

TEACHERS' CONCEPTUAL UNDERSTANDING IN SCIENCE: NEEDS AND POSSIBILITIES IN THE PRIMARY PHASE

Terry Russell, Derek Bell, Linda McGuigan, Anne Qualter, John Quinn and Mike Schilling

Centre for Research in Primary Science and Technology (CRIPSAT), The University of Liverpool, Dept. of Education, 126 Mount Pleasant, P.O. Box 147, Liverpool L69 3BX, UK

Abstract   This paper reviews the changing expectations made of primary teachers in respect of their science subject-matter knowledge and understanding over the past 20–30 years up to the present time. The shift from a process orientation to a post National Curriculum emphasis on teachers' science knowledge and understanding is discussed. While a balanced approach which takes account of science processes, content and contexts is advocated, it is acknowledged that available evidence points irrefutably to the advantages of enhanced subject matter knowledge on the part of teachers. It is argued that the case for teachers' subject knowledge can be overstated, including a logical fallacy that this necessary knowledge also equips teachers to teach science in primary classrooms. A range of CRIPSAT projects which throw light on the debate are reviewed, including ways of making science knowledge accessible to teachers in a form which also addresses their classroom needs. In particular, it is suggested that a greater emphasis on teachers' own science knowledge is not incompatible with a constructivist, child-centred approach to science education in primary schools.

Introduction

Within a generation, primary science education has been transformed; the developers have moved into what used to be a backwater. Only 14 years ago, Harlen (1978) was able to describe the situation in England as one 'where the ethos of primary education is child-centred and the decentralised system gives teachers freedom and responsibility for making decisions about children's learning' so that, as a consequence, 'the conditions favour a local choice of content'. Currently, the balance of political and professional contributions (both of which must always be present and always interact) to the debate about the direction practice should take are a matter of sensitivity. The term 'child-centred' is now used by politicians as a term of abuse; the content is centralised; what is expected of teachers has changed radically. Expectations about how teachers should teach and what they should teach have changed; assumptions about what science teachers should themselves 'know and understand' have a central place in the debate. This paper reviews the reflections and experiences of one group of researchers with particular regard to the issue of primary teachers' conceptual understanding in science.

Two main issues serve to focus the discussion:

(i) a consideration of the needs of teachers in terms of their own understanding of science concepts, in order to be considered effective deliverers of the primary science curriculum.

(ii) a review of the possibilities for making such understanding acccessible to teachers through both pre-service and in-service training experiences.

A range of evidence is drawn upon, starting with an historical review of expectations, moving on to more recent and direct experiences of the authors, including implications drawn from several recent research, development and curriculum projects based at the Centre for Research in Primary Science and Technology (CRIPSAT). Although the evidence is incomplete, some of the issues which are in need of further debate and research are identified.

Perspectives on Teachers' Knowledge and Understanding for Primary Science

The backwater which used to be primary science was broadly characterised as 'nature study'; though it might not always have been referred to as 'science', the liberal ideal of exciting children's interest in the world around them governed this curriculum area. The tradition is a respectable one and there is an obvious evolution into more topical environmental concerns. The knowledge and understanding which was required of teachers and pupils alike might be described as 'general knowledge' rather than science-specific, while the higher order science process skills—hypothesising, planning investigations, interpreting data—would not normally have been called upon by most teachers operating in the nature study framework. Harlen (1978) quoted the Primary Schools Science Committee of the ASE:

> At this level we are concerned more with the development of an enquiring attitude of mind than with learning facts. (ASE, 1963)

Piagetian research, replicated in the UK, was one important influence which began to have an impact on primary mathematics education in the 1960s and 1970s (for example, Peel, 1960); to a lesser extent, there was an impact on primary science. Particularly, there was interest in describing modes of cognitive functioning, thinking 'stages' and *matching* curriculum experiences to children's developmentally defined needs. Whether the stance adopted was Piagetian (learning occurs in an invariant developmental sequence) or Brunerian (teach anything, any time, by making it accessible), the novelty of the approach resided in the focus on the learning *process*.

This developmental perspective applied to learning across the primary curriculum. The theoretical orientation carried an intrinsic appeal in being congruent with teachers' perceptions of children as active learners, each at different points in cognitive and language development and with differing social and emotional backgrounds and needs. The orientation is caricatured in the slogan, 'we teach children, not subjects'. Hyperbole apart, herein lie the origins of the 'child-centred' climate in primary classrooms which is currently being challenged. The current political exhortation is for the teaching of subject-matter. The ramifications of this policy position are apparent in assessment (the retreat from science process assessment in Key Stage 1 SATs) and teacher training (licensed teachers) (NCC, 1991).

Throughout the 1970s and 1980s, the process orientation to science education was promoted and was most visible in primary classrooms. The 'Assessment Framework' developed by the APU science assessment team (see, for example, Russell *et al.*, 1988) reflected this orientation; 'Applying Concepts' was represented as one part of one of six process categories. Despite the process orientation, an important insight of the APU science programme that has possibly received insufficient emphasis is that the cognitive burden of a science task results from an *interaction* of *process, concept and context*. (See for example, Kirkham, 1989). It was recognised that all these factors have an impact on the practical science activities which teachers present to children. Although constructed for assessment purposes, the APU framework had a significant bow-wave effect on curriculum practice (as assessment structures inevitably do). The message to teachers was, '*Approach science content through the process skills*'.

The View from HMI

It is interesting to review how the traditionally 'impartial and independent' HMI have viewed the practice of primary science over the past decade and more. In their report of Primary Education in England (DES, 1978) they stated that:

> The most severe obstacle to the improvement of science in the primary school is that many existing teachers lack a working knowledge of elementary science appropriate to children of this age. (para 5.83)

This sentiment was echoed by HMI in an overview of primary science (DES, 1989) when they commented:

> Promising developments in science teaching in primary schools over the decade (since 1978) underline the importance of two conditions for improving the quality of work. Opportunities are needed for teachers to:
>
> > improve their own knowledge of science in relation to what the children should know, understand and be able to do;
> >
> > observe work of good quality taught by more experienced colleagues and discuss how it is planned, organised and sustained. (para 100)

Teachers' own knowledge of appropriate science has been on record as a cause for concern amongst HMI for many years, though it is important to add, only as one issue amongst many. HMI reported (DES, 1989) that:

> The amount of science taught has increased markedly . . . (para 4)
> . . . (there has been an) improvement in teaching competence and in the confidence of teachers to provide more physical science (para 5).

Bennett & Carré (1991) reported:

> . . . an increased feeling of competence by teachers themselves.

HMI (1985) in the DES policy statement (para 25) foresaw that 'teachers will need help and support' in helping children to 'gain a progressively deeper understanding of some of the central concepts of science'. The knowledge and understanding required is not general knowledge or 'common-sense', but very specific and particular knowledge of scientific concepts; it is not the knowledge that every individual picks up along the way, for survival or out of casual interest.

In many instances, scientific knowledge is counter-intuitive, contrary to common-sense. Much of the adult population, including that proportion taking up teaching as a profession, has had minimal exposure to science. Research very soon confirmed the implication of this lack of science background in the primary teaching profession: many teachers revealed very little conceptual understanding in science. Even worse, but true to constructivist predictions, they had nevertheless attempted to make sense of the phenomena they encountered and many revealed the same explanatory inadequacies (or 'misconceptions') as the pupils whose scientific understanding they were charged with shaping. (Kruger & Summers, 1988). In a sense, teachers were sitting ducks, a soft target for this kind of research.

Bennett & Carré (1991) specified the advantages to teachers of possessing subject-matter knowledge and understanding:

> . . . teachers need such knowledge adequately to transform programmes of study and attainment targets into worthwhile and appropriate tasks, they need it to frame accurate and high quality explanations, and they need it to diagnose accurately children's understandings and misconceptions.

The same authors have also reported initial results on the impact of subject knowledge on teaching performance. Available analyses refer to music only. On the basis of observational ratings the suggestion was:

> In comparing music lessons given by specialists with those given by non-specialists it is very clear that not only do the specialists maintain a better balance between teaching and management, they do so at consistently higher levels of performance.

This kind of analytical and empirical contribution to the debate is most welcome. An extraordinary feature associated with primary teachers' predictable lack of scientific understanding has been the manner in which there has been, from some quarters, an explicit attribution of blame. This phenomenon is not unique to the UK. Seddon (1991) commented:

> Recent Australian reports on teachers and teaching suggest that teachers are a 'problem' and on this basis make a range of reform proposals for teacher education and in-service work. . . It implies that the problem of teachers and teaching is one of individual deficit rather than contextual limitations.

Teachers have been criticised for lacking the knowledge which a radical change in their job definition requires them to have. In time—towards the end of the first decade of the twenty-first century—those school leavers who decide to train as teachers will themselves have been through a basic schooling in science as a core subject. These teacher trainees will not have had the option of dropping science. (The *quality* of the science teaching they will have received is less certain.) Within a generation, this newly acquired expertise will have permeated the teaching profession. However, radical changes in the educational system are not supported by such long-term evolutionary changes. Pre-service and in-service training needs are more immediate. In the context of the radical changes which have since become law, equally radical changes in practice are required of teachers; to describe these new requirements as 'demanding' is an understatement.

Pre-Service Training Implications

There are indications of radical policy changes in teacher training. McNamara (1991) suggested:

> . . . there can be little doubt that at present promoting the importance of subject matter is seen as the way to reform teaching training and foster the quality of learning in schools.

The establishment of the Council for the Accreditation of Teacher Education (CATE) in 1984 contributed to the trend of greater subject-based preparation for teachers in training. Pre-service training for primary teachers has seen the requirement for the teaching in Mathematics and in English quantified (DES, 1984) and later, equivalent provision in science has been added to the formula. Teacher-trainees are required to have a recognised proficiency in English and Mathematics; will the same follow for science? It seem illogical that it should not be so.

A minimum of two years academic subject study is now required as part of a four year course of teacher training. Students entering postgraduate courses are assumed to possess the requisite knowledge but their first degree must be in an approved subject area. Both of these requirements appear to be based on the assumption that if the trainee teacher has knowledge in depth in one or two subject areas then this will provide for more effective teaching and learning.

Currently, there is no National Curriculum in Initial Teacher Education. In subject terms, however, it would seem to make sense that such a curriculum would comprise that part of the National Curriculum (5–16) which would enable every teacher to be at a level of understanding, knowledge and competence that is further advanced than his or her students. This standard might be the minimum requirement to enable the teacher to plan for progression in learning.

Initial Teacher Education offers the possibility of a fresh start within whatever ground rules apply. This is not to deny the difficulties which clearly exist in meeting policy expectations. For practising teachers, the rules have changed while the game is in progress and it is much more difficult to describe an overall picture of strategies and possibilites in relation to supporting teachers' own conceptual understanding. A number of insights do emerge from several CRIPSAT projects and initiatives, and it is to these that this paper now turns. The activities to which the following sections refer are:

    In-service Education for Primary Science
    The Primary Science Teaching Action Research Project
    The Primary Science Processes and Concept Exploration Project
    NCC Distance Learning Material
    Telephone Help Line Support

In-Service Education for Primary Science

In 1985, LEAs bid for money through the Educational Support Grant (ESG) scheme to support teaching science in primary schools. The IPSE evaluation (ASE, 1988) identified a number of features of the in-service training and school-based support that had been offered, indicating that the emphasis was usually on process approaches. Little time had been given to considering the underlying scientific concepts. (See, for example Kinder & Harland, 1991).

Concurrently, DES circulars described funds available to LEAs to second teachers to attend long courses. Designed for primary science co-ordinators the focus of these 35-day courses tended to be not *what* was learned in science but *the way* it was organised and taught. Developing the skills of a curriculum co-ordinator was also regarded as important in some of these DES-approved courses. Such investment by a school was likely to have a payoff in terms of whole-school development, using the seconded teacher as the agent of change.

More recent funding has been aimed at helping teachers to implement the National Curriculum. The DES had decided that courses in 'Physics for Primary Teachers' were needed; this idea evolved through consultation into the current 20-day DES-designated courses, 'Science for Primary Teachers'. For a small number of teachers, 20-day secondments are available in LEAs which have negotiated with an institution of HE and secured DES funding for a proportion of the costs. The guidelines were explicit that the courses should be geared to *improving teachers' knowledge of science*; physical science was to have a high profile. The aim was to equip participants with the subject knowledge and understanding which would enable them to teach all the Attainment Targets in Key Stages 1 and 2.

The negotiations preceding eventual DES approval of CRIPSAT's 20-day programme clarified the official requirements: the course should focus exclusively on background knowledge, linked closely to the National Curriculum—in particular those areas which the DES had identified as important for teachers. Nevertheless, the mode of delivery of a body of knowledge to teachers will carry implicit messages as to the institution's views about how effective learning occurs. To utilise science process skills as the means of supporting the development of teachers' understanding is to make manifest important assumptions about their role in effective learning. The principles of active learning might be well appreciated by an adult learner, but the practice of translation into active learning opportunities for children would have to be the subject of another, classroom-focused course of study.

About 60 teachers from four LEAs participated in the approved courses run by CRIPSAT during 1990 and 1991. A constructivist approach was adopted in the efforts to facilitate the teachers' learning in science, this being most consonant with the providers' experiences and beliefs about effective learning. The work of the primary Science Processes and Concept Exploration (SPACE) project was examined in some detail, with the advantage of much of the research work having been conducted 'on the doorstep'. (It was noticeable that a number of course leaders from other institutions were ordering multiple copies of the SPACE project research reports for use on their own 20-day courses; these publications describe research into *children's* ideas. See Osborne *et al.*, 1990, for example).

It was soon realised that 20 days (120 hours) is simply not enough to cover the whole of the science curriculum, assuming all domains need attention. Some Attainment Targets therefore received less attention than others.

The courses had been designed to encourage teachers to explore their existing knowledge by describing their understanding of events and phenomena. Discussion with an 'expert' has, for many of the teachers, resulted in the realisation that learning in science can be exhilarating. This has contrasted with their previous experience of learning in science: it was usually not about learning, only about being taught, in their own secondary schooling.

Not all participants' expectations were met. Some felt that the set of facts which they anticipated taking away and forwarding to the next set of recipients had not been handed over. Perhaps such courses result in two different sorts of product: a set of teachers who rediscover the joy of learning at their own level—and who finish the course inspired to find out more science; another set who might have been able to select a few facts from the explorations and discussion which they could share, possibly appropriately, with children in their class. *In neither case was there any certainty that useful changes in teachers' ability to manage meaningful learning experiences in the classroom had been affected.* That issue simply had not been addressed. There is need for an enquiry which studies the changes in teachers' practice as a result of their experiences of having their ideas challenged and changed.

The Primary Science Teaching Action Research Project

The idea that children's learning of science content might be approached through science processes assumes teachers to have an understanding of those processes and the possibilities of classroom management and organisation which encourages their deployment by pupils. The Science Teaching Action Research (STAR) Project (Schilling *et al.*, 1990; Cavendish *et al.*, 1990; Russell & Harlen, 1990) focused on the classroom development of process skills through action-research. Observers recorded the incidence of a range of process skills occurring in interactions during science lessons. Children's use of process skills was also assessed through pencil and paper and practical modes. It emerged that the more general process skills—observation, communication, recording, for example—were more in evidence, while some of the more science-specific skills—hypothesising, framing investigable questions, critical reflection, repeated measures—were scarcely in evidence. Although systematic data were not reported with regard to lesson content, it was often apparent to classroom observers that the science *content* was of a trivial nature. To some extent, this was understandable in that the project teachers were specifically attempting to develop the process dimension of their lessons. But evidence of process science lacking any substantial science content might be regarded as a cause for concern, and a good reason for examining more closely the content agenda of primary science.

Teachers' Knowledge and Understanding of Science and the Primary SPACE Project

The Primary Science Processes and Concept Exploration (SPACE) Project began in 1986, operating jointly through CRIPSAT and the Centre for Educational Studies at King's College, London. When the project was conceived, there was no National Curriculum, though the context of thinking on curriculum content has been referred to above.

The first three years of the project concentrated on collaborative programmes of research, with teachers, into children's understanding and possibilities of helping children to develop that understanding. The orientation was constructivist; the pragmatics of working with teachers ensured that the techniques developed were viable in normal primary classrooms. The project then moved on to develop curriculum materials which built on research findings. Both the research phase and subsequent trial of curriculum materials offer insights into teachers' needs in terms of their understanding of the science which they were attempting to teach.

The collaborative nature of the project included regular meetings with groups of teachers during which concerns could be expressed. Although the systematic collection and reporting of data focused on children's ideas and changes in these as the result of classroom interventions, records of the views expressed by teachers were also maintained. It is these records which inform the following comments.

The teachers who were involved in the SPACE project were not selected on the basis of any particular competence in science. They varied in their own experiences of science, as well as in their approach to primary science teaching. Many were hesitant about their own understanding of science. For some teachers uncertainty was restricted to particular concept areas; other referred to their own level of understanding of science as a whole. Some reported a fear of transmitting wrong ideas and causing pupils 'irreparable damage'.

Teachers also mentioned that they did not know how to teach science in a way that they believed would fit in with their conceptions of good primary practice. Many had negative feelings about their own experience of learning in science and wanted to ensure children understood and enjoyed the science they were offered. Many teachers were concerned that they didn't know how children's understanding in science might develop and how far it was appropriate to try to take children within a particular concept area.

The problems teachers were facing were rooted in an image of themselves as not knowing; not knowing how to plan science, not knowing how to teach science, and not knowing and understanding the science conceptually. Inevitably these responses to science affected classroom pratice. Some teachers were reluctant to engage with the science concepts, relying exclusively on the development of process skills (as described above in the context of the STAR project). Others adopted a didactic approach, transmitting specific facts; in these lessons there would be little deviation from the lesson plan. This strategy would ensure there was no confusion in the message transmitted and avoided any discussion in which the teacher's own understanding might be challenged. The adoption of this didactic teaching style by teachers teaching an unfamiliar science was also observed by Cronin-Jones (1991). Several teachers reported that they were reluctant to embark on practical science because they didn't know where to start; others described lessons as consisting of isolated, unrelated science investigations and a few explained that lessons were dominated by worksheets.

Although the project primarily addressed conceptual understanding, teachers were equally anxious about how they might support the development of children's understanding by encouraging the use of process skills. When children expressed certain ideas, teachers were unsure how to help children to do anything with those ideas—in particular, how to operationalise ideas in a form which could then be investigated.

In addressing doubts about conceptual understanding, sessions during which teachers attempted to articulate their own understandings of various ideas to one another were found to be very helpful. Such sessions were centred on simple practical experiences, just enough practical activity to set a context, with the main value of the experience residing in the attempt to express personal understanding *and* to understand another individual's attempts to make their viewpoint intelligible. As well as being a valuable cognitive or intellectual rehearsal for subsequent teaching activities, such interactions also brought to the surface feelings of vulnerability and on occasions, a sense of inadequacy concerning science knowledge. Sharing this sense was useful therapeutically, in that all could identify with such feelings; group cohesion and mutual support in getting to grips with the science knowledge often resulted, together with a gains in confidence.

The current debate about primary teachers' knowledge and understanding tends to have neglected the issue of teachers *supporting one another in developing knowledge and understanding*. The SPACE meetings enabled teachers to begin to appreciate the possibilities of learning in science as an interactive rather than isolated activity and group sessions appear to have contributed significantly to the development in teachers' understanding. They began to test their ideas, to consult secondary sources, to reflect on each other's ideas and modify their thinking on the basis of their discussions. Kruger & Summers (1989) also reported teachers modifying their ideas as they reflected on their explanations. The meetings also provided insights into the intellectual and emotional struggle which their pupils would be experiencing during science lessons, just as they had themselves.

Meetings also helped teachers to appreciate, as they considered changes in scientific understanding (such as is seen historically in, for example, the classification of plants), that science is not a fixed body of knowledge. This outlook made it easier to consider children's ideas as acceptable, in a provisional and developmental sense. The meetings encouraged teachers to engage with the science and plan possible starting points which could lead to a development in children's understanding. One of the consequences was a reported increased confidence in the ability to plan appropriate starting points:

> I am more confident about starting somewhere with the children, and getting them involved, then to direct them . . . before I just jumped in blindly.

> I can put my knowledge into a framework to use with the children at their level.

> Now I'll have a go, I feel a lot clearer in my mind where I'm going and why I'm doing things.

Effective planning included anticipating the ways in which children's thinking might progress. Lessons were no longer seen as a separate, unrelated series of investigations, but part of a coherent plan. Teachers felt that they were more able to provide a differentiated science curriculum which matched the individual needs of the children:

> It does not depend on me deciding at what level the children will understand a certain concept. The process starts from them and I find it easier to then help the children to progress rather than imposing a level on them that is inappropriate and I lose half the class.

Teachers mentioned that confidence in their own science understanding had improved along with their teaching. The interactive nature of the approach, both one to one and as a whole class, involved all participants in some reflection on their own ideas in the light of ideas expressed by others. Teachers reported that the ideas and questions emerging from children challenged their own undertanding of the background science. It seems that eliciting children's ideas also serves to confront teachers' own understanding and activates an awareness of the limitations in their own explanations. This recognition may prompt them to reflect further on their own ideas:

> Elicitation of children's ideas certainly makes me reflect on my own understanding. Realising that children have ideas similar to my own encourages me to pursue my own understanding, and this in turn helps me to better guide the children's understanding.

The main thrust of the second—curriculum development—phase of the project was to develop materials for teachers which would support the elicitation of children's initial ideas before helping the process of conceptual development. (Nuffield Primary Science—in press). It was recognised that teachers would occasionally find themselves lacking basic factual knowledge, even when they pooled their knowledge resources within a school. A concise section, 'Background Science', was included in each of the Teachers' Guides. The language used was as straight-forward as possible, supported by diagrams. The intention was to set out basic principles for reference purposes. These sections were well received; some teachers reported that even when they were not used, they provided a sense of emergency support, like a safety-net, which can be ignored once its existence is established.

Overall, what the research and curriculum development activities of the SPACE project revealed is that teachers' knowledge and understanding needs are just one facet within a programme of curriculum reform:

> I had positive feelings towards science but was very disillusioned as to the way it was taught in school. I did not seem to find any materials which were not text book or were worksheet style and I did not feel this was appropriate.

Teachers were clear that they needed materials which helped them with their background knowledge, provided them with starting points for practical investigative science, and enabled them to plan for the development in children's understanding. The materials also needed to be useful in a variety of organisational settings and flexible enough to be incorporated into teachers' own topic plans.

Distance Learning Material

Another recent project in which CRIPSAT was centrally involved was the development of distance-learning material aimed at helping teachers to develop their background knowledge and understanding in science. *'Knowledge and Understanding Science: Forces'* (Schilling *et al.*, 1992) was the first in a proposed series of books commissioned by the National Curriculum Council. Some have questioned whether the development of textual materials to support teachers' understanding is consistent with the avowed constructivist philosophy of the authors. Indeed, the same question has been raised in relation to the Background Science sections of the SPACE material. The answer is multi-faceted, just as learning is. Firstly, adults can be assumed to share many of the characteristics of learning of children, so a constructivist approach would be appropriate. But a constructivist view of adult learning must be able to accommodate learning through *all* media, including more abstract media which are not as accessible to children simply because children do not have the experiences to draw upon which can link text to life and make it meaningful. Secondly, to some extent, even textual forms can adopt an interactive approach with the reader in which reflection on beliefs as well as direct experiences can be encouraged. This interactive approach was the one adopted. The material was also designed to include a 'fast-track' for those who wanted only a rapid review of the science as a scientist might express it. This was provided in sections headed, 'What the Scientists Say'. The use of the plural contains a hidden irony for the authors, for establishing consensus between a group of physicists on the subject of 'Forces' proved to involve considerable negotiation and compromise. Those who advocate a return to 'simple, straight-forward factual information' might have been shaken from their innocent delusions had they been involved in the writing process. The science that was eventually presented was not the 'last word', for there is no such thing in science. It was science written at a particular explanatory level for a particular group having a particular use for it.

The significant omission—as that is how it can be perceived—is any reference to classroom transactions. The reader of the book (as the learner on the 20-day science course) must appreciate and endeavour to internalise the active and interrogative process of learning through which they are themselves proceeding. What is attempted is demonstrable learning by teachers and *at the same time* explicit strategies used in facilitating that learning. It is not that the content is suddenly important at the expense of the process; the teachers have to use (and we have to encourage) the aspects of science processes which are appropriate for experiential learning, even at a distance. This can be achieved if teachers can be encouraged to reflect on the process of their own learning as they work through the materials.

There are compromises. The defined amount of knowledge and understanding to be covered—Old AT10; levels 1 to 7—determined the minimum length of the pack. The material was also written to accommodate what came to be known as the 'worst case scenario'—the teacher totally isolated from any other source of support, from colleagues, advisers, etc. Such total isolation should be a fantasy; discussions about distance learning rapidly turn to a consideration of ingenious strategies for bringing people together, either directly or via electronic means. Not least, it has to be acknowledged that this kind of material will work only for those who want it to work for them; they must be motivated to meet the material half-way. The NCC is wise to have decided on a strategy of making such material available to those who *request* it.

Telephone Help Line Support

Not all teachers gain access to courses. In June 1991, the General Secretary of ASE invited offers of support for a primary science telephone help line (ASE, 1991). The proposal from CRIPSAT was accepted and the Help Line has served primary teachers in four LEAs for a trial period of three months, to date.

The Help Line was initially conceived as providing an accessible source of science information to the teacher who perceived such a need. In setting up the service, care was taken to point out that a telephone link could not provide a substitute for the personal contact and advice available through LEA advisers or advisory teachers (where these are available). For example, matters of school or authority policy would require local and personal contact.

How exactly the resource might be used in practice was a matter of conjecture. To cater for the factual requests which teachers might make, various sources of information have been assembled. These include compilations of useful articles, background science notes, ideas for classroom activities, equipment, resources, together with ideas for visits and other services provided locally by museums, etc.

While the Help Line has to cater for brief factual requests for information— 'How old is the Earth?' was the first question received—it is hoped that the link might also be able to support a more reflective approach to primary science and a more interactive approach to enquiries. Even if it were technologically possible, an automated 'speaking clock' kind of response to enquiries is not favoured. What has been promoted so far is a more thoughtful approach to the development of teachers' understanding. It is assumed that only a small sub-set of subject-matter knowledge is appropriately transferred directly from textbook or database to the classroom. The quality of the science teaching and learning occurring at the other end of the telephone line is a matter of interest.

The nature of the problems which teachers feel can be approached usefully via a telephone service will emerge as the service is used. The extent to which it is possible to go beyond the provision of factual information towards supporting understanding will be a matter of interest. It may be that the relative anonymity of the medium will make teachers feel less vulnerable about not knowing, more prepared to engage in dialogue about what they do not understand.

It is intended to record just how the service is used; this will provide information about the areas of the curriculum in which teachers are most likely to seek the kind of support offered. The nature of any documentary support requested or offered successfully will also be of interest, as will the incidental insights which are likely to emerge. It seems likely that requests for help will be encountered which cannot be addressed by the telephone medium or follow-up mailing. To know that such support is being requested will enable alternative support strategies to be framed.

Conclusions

The argument that teachers should have a certain basic level of understanding of the science they teach is irrefutable. As the curriculum has changed, so that basic need has changed. In the days of Nature Study, a general knowledge supplemented by Field Guides and a few basic reference works on the shelf might have sufficed. As science education in primary schools moved to an emphasis on process science, delivered only by those who had the confidence to do so since the subject was not a mandatory requirement on the primary curriculum, it was justifiable to encourage teachers with their attempts at an active, investigatory science. This suited the ethos of most primary classrooms and there was faith that the science knowledge would gradually be acquired.

The more recent shift has been to a subject based on enquiry, investigation, and knowledge and understanding. For the majority of primary teachers the shift came with the introduction of the National Curriculum with its emphasis on whole investigations and high content burden.

The argument that a certain amount of science knowledge and understanding is *necessary* to teach the current curriculum is accepted. However, there is logical fallacy in the overstated version of the same case which attempts to suggest that such knowledge and understanding is *sufficient.*

The shift in emphasis towards the content of science has been partly the result of political pressure to move away from a child-centred experiential approach to teaching and towards the adoption of more formal methods. The 'argument' against a child-centred approach is that it is 'trendy', which might be inferred to mean superficial, fashionable, ephemeral. While the term itself may have been obscured beyond worthwhile redemption, some of the implications for the practice of teaching science must be strenuously defended. The essential element is the importance of children's starting points, the initial attitudes, beliefs and ideas which they bring with them to their science lessons, are of critical importance to teaching and learning. There is now overwhelming international support for such a view. (Pfundt & Duit, 1991, cite 2,000 bibliographic references to work in this area, confirming the enormous interest and commitment of the science education community to the approach). What does need to be unravelled is the nature of science teaching practice which is informed both by a constructivist (child-centred) view of learning and an acknowledgement of the importance of teachers' own subject-matter knowledge and understanding. This is the critical issue confronting teachers and researchers.

It has to be accepted that children bring a whole range of strongly held, intuitively based, common-sense ideas with them to their science lessons. Another set of interpretations—the 'official' set— is laid out in the Order for science. It is irrefutable that teachers must share the knowledge and understanding of science set out in the National Curriculum. But having the knowledge is not enough, never has been and never will be; the advocacy of subject-matter centred didacticism ignores the constructivist evidence. Children *interpret* their world; part of the world they interpret is that which impinges on them in school. They *interpret* the evidence, the proofs, the facts, which teachers put in front of them. This is the essence of being a human, active learner. It appears that children and adults alike are 'hard-wired' to behave in this way—'Man, the Scientist', as George Kelly put it (Kelly, 1963).

The political imperative has influenced the opportunities which have been available for providing INSET for teachers. It seems that teaching teachers science may meet only part of their needs, and that without providing the support they need to develop teaching strategies the courses and learning materials which are now being funded may be failing the recipients. Present policy initiatives focus on the science knowledge without paying heed to the *manner* in which that knowledge is acquired, nor how *effectively* it is acquired, nor what teachers might do with it *in classrooms*, vis à vis children, once they have acquired it.

The development of in-service materials which provide only the science knowledge run the risk of reinforcing some of the teachers' perceptions of science as remote and difficult to understand. The lesson from the Primary SPACE Project seems to be that materials which engage teachers' thinking are those which encourage them to reflect on their own understanding, together with children's understanding, and on their classroom practice.

The reactions of teachers suggest a number of ways in which the SPACE project has helped them to improve their science teaching. Difficulties with personal understanding seem to be interwoven with additional problems associated with low levels of confidence. Teachers' failure to understand science emerges as both a cognitive and affective problem. A case can be made that their difficulties will not be resolved by attempts to provide *only* the science knowledge. Indeed, many who described themselves as having sufficient knowledge for the level they had to teach, or who mentioned that they had some science qualifications, still expressed doubts about their ability to teach science effectively. Their difficulties lay in developing an approach to science teaching which they accepted as good primary practice, an approach which offers young children the scientific experiences they need, an approach which takes account of children's current level of understanding and which enables teachers to identify how chidren's thinking might progress.

Another unnecessary assumption is that every individual has to proceed towards a position of understanding as though no other source of support is available. School-based group activity can be a much more acceptable way of proceeding, provided there are basic frameworks to which groups can refer. It is to be hoped that the distance learning material desribed above will be used in this way.

Is it necessary to go as far as calling for subject specialists in primary schools? The science knowledge required is surely not so great that it is not within the range of every primary teacher, *given the means and the opportunity*. What, after all, is the minimum knowledge and understanding of science that every teacher must have? As much as is required of the children they teach? A little more— perhaps a key stage more—than their pupils? The needs can be overstated, but a clear sight of what teachers are required to teach and what science they themselves need to teach it, is needed.

Is it legitimate to move from the idea of *informing* to the notion of providing *access and support*? That is, recognising the impossibility of instilling encyclopedic science knowledge in every teacher's head, and instead, attempting to provide an accessible source of information? There is a need for caution here, for it is too easy to slip into an upside-down logic, to re-define problems in terms of available solutions. It would be very convenient if, by readily accessible electronic means, science teaching problems could be solved. It would be even more convenient if pupils could assimilate all the science educators could wish them to have as the result of exposure to flawless electronic media. This is not the way that teaching and learning happens. Enough is known about learning to know that there is much still to be learned; positions of ideological purity would be folly, but so too would be over-stated and simplistic solutions. Primary teachers find themselves in a very exposed and vulnerable position with regard to their conceptual understanding in science. The range of backgrounds and differences in experience and prior knowledge call for an imaginative range of options to provide the support needed. No single option should be hailed as the panacea; no strategy should be dismissed as the 'sticking-plaster'. Education, for teachers and for children, remains the art of the possible.

References

Association for Science Education (1963) *Policy Statement prepared by the Primary Schools Science Committee*. Hatfield ASE.

——(1988) *Initiatives in Primary Science: an Evaluation*. Report, Hatfield: ASE.

——(1991) *Education in Science* 143, 5 June 1991.

Bennett, N. and Carre, C. (1991) No substitutes for a base of knowledge. *Times Educational Supplement* November.

Cavendish, S., Galton, M., Hargreaves, L. and Harlen, W. (1990) *Assessing Science in the Primary Classroom: Observing Activities*. London: Paul Chapman Publishing.

Cronin-Jones, L. (1991) Science teacher beliefs and their influences on curriculum implementation— Two case studies. *Journal of Science Teaching* 28, 3 (March).

DES (1978), *Primary Education in England: A Survey by HMI*. London: HMSO.

——(1984) *Initial Teacher Training: Approval of Courses*, Circular 3/84. London: HMSO.

——(1985) *Science 5–16: A Statement of Policy*. London: HMSO.

——(1989) *Aspects of Primary Education: The Teaching and Learning of Science*. London: HMSO.

Harlen, W. (1978) Does content matter in primary science? *School Science Review* 59, 614–25.

Kelly, G. (1963) *A Theory of Personality. The Psychology of Personal Constructs*. New York: W. W. Norton and Co.

Kinder, K. and Harland, J. (1991) *The Impact of INSET: The Case of Primary Science*. National Foundation for Educational Research.

Kirkham, J. (1989) Balanced science: Equilibrium between context, process and content. In J. Wellington (ed.) *Skills and Processes in Science Education. A Critical Analysis*. London: Routledge.

Kruger, C. and Summers, M. (1988) *Primary School Teachers' Understanding of Science Concepts— General Overview, Working Paper No. 1, PSTS Project*. Oxford: Oxford University Department of Educational Studies.

——(1989) An investigation of some primary teachers understanding of changes in materials. *School Science Review* 71 (255).

McNamara, D. (1991) Subject knowledge and its application: Problems and possibilities for teacher educators. *Journal of Education for Teaching* 17 (2), 113–28.

National Curriculum Council (1991) *The National Curriculum and the Initial Training of Student, Articled and Licensed Teachers*. York: NCC.

Osborne, J., Black, P., Smith, M. and Meadows, J. (1990) *Light. Primary SPACE Project Research Report*. Liverpool: Liverpool University Press.

Peel, E. A. (1960) *The Pupil's Thinking*. London: Oldbourne Book Co. Ltd.

Pfundt, H. and Duit, R. (1991) *Bibliography. Students Alternative Frameworks and Science Education (3rd Edition)*. IPN Reports in Brief. Kiel: University of Kiel.

Russell, T., Black, P., Harlen, W., Johnson, S. and Palacio, D. (1988) *Science at Age 11. A Review of APU Survey Findings 1980–84*. London: HMSO.

Russell, T. and Harlen, W. (1990) *Assessing Science in the Primary Classroom: Practical Tasks*. London: Paul Chapman Publishing.

Schilling, M., Harlen, W., Hargreaves, L. and Russell, T. (1990) *Assessing Science in the Primary Classroom: Written Tasks*. London: Paul Chapman Publishing.

Schilling, M., Atkinson, H., Boyes, E., Qualter, A. and Russell, T. (1992) *Knowledge and Understanding of Science: Forces: A Guide for Teachers*. York: National Curriculum Council.

Seddon, T. (1991) Rethinking teachers and teacher education in science. *Studies in Science Education* 19, 95–117.

RAISING AND ANSWERING QUESTIONS IN PRIMARY SCIENCE: SOME CONSIDERATIONS

Catherine Woodward

University College of Swansea, University of Wales, Department of Education, Hendrefoilan, Swansea SA2 7NB, UK

Abstract In England and Wales, teachers of primary science are required by statutory orders, specified in the current National Curriculum documentation, to encourage children's questioning skills. In this article the educational significance of developing interrogative skills in science is explored alongside a consideration of current classroom practice. The apparent difficulties of implementing this aspect of the statutory requirements are examined with reference to studies in the field of questioning. Emerging from the discussions are issues reflecting the need for practitioners to be supported in the review and evaluation of pedagogy as a way of providing professional development concurrent with improved application of questioning strategies.

True learning is characterised not so much by the answering of questions as by the asking of them. (UNESCO)

Current documentation referring to National Curriculum requirements in science education in England and Wales states the need to plan activities which 'encourage the raising and answering of questions'. (DES, 1991; NCC, 1991). A further development of this aim is the ability to 'ask questions. . .in a form which can be investigated'.

In this brief paper it is the intention to consider the place of questioning in the teaching and learning of science at the primary stage. It is proposed that further attention should be given to this skill area so that its potential to enhance children's involvement in scientific activity is further exploited.

Discussion of the questioning process and its implications will fall into five main sections as follows. First, it is necessary to consider the rationale for promoting pupils' questions and highlighting any advantages that might be accrued from developing questioning skills. This will be followed by a consideration of the effect of teachers' questioning of pupils with a brief reference to literature in this area. Having identified the possible gains from pupil questioning, we will then examine the 'state of the art'. How prevalent are pupils' questions in classroom practice of primary science? Are pupils' questions actively encouraged and do such questions lead to enhanced involvement in science activity?

Following on from a consideration of present practice, it is apt to take a closer look at possible reasons for pupil questioning being a problematic area and attempt to analyse the difficulties teachers face in implementing the National Curriculum requirements related to questioning. Finally, some suggestions are proposed for positive ways in which the increased involvement of pupils in questioning behaviour can become a reality in the classroom.

Why Promote Children's Questioning?

It is reasonable to propose that one of the aims of primary education is to support and encourage children in their development as independent learners and critical thinkers. The ways in which teachers can promote the notion of independence in learning include the provision of opportunities for children to become questioners, the encouragement of pupils to enquire and seek information, and a recognition of the role which children's questions can play in fostering learning.

In order for learning to occur, it is, of course, not a prerequisite that questions raised by children must be of a kind which can be investigated by scientific activity. On the contrary, much learning can arise from various categories of questions whether or not they lend themselves to an exploratory activity. By posing questions, children are learning to apply language in a specific way which demands the use of particular linguistic strategies. They are also learning that asking questions is an accepted way of finding out more about things that interest or puzzle them. However, in order for children to raise questions appropriate for scientific investigations it is necessary to assist them in moving from general questioning to recognise those questions which *can* be pursued by their classroom activity. This process is a gradual one as Harlen (1985:34) acknowledges:

> To develop this awareness is a significant part of their education, but it will come only very slowly. . .

An alternative perspective for promoting children's questioning is offered by White (1977:125) who suggests that schools should nurture in children 'the habit of contemplation' in order to enrich life. The means for doing so, he proposes, rests largely with 'asking one's self questions and hazarding answers'. Furthermore, White sees an analogy between asking questions and creativity claiming that:

> If the student never thinks of any questions he is not going to be creative.

The relevance of scientific questioning preempting inquiry in adulthood is not a new phenomenon. Sumner (1947) writing in a handbook on the teaching of science again draws our attention to the need to develop enquiring minds:

> The intellectual interest in scientific discovery for its own sake is worth fostering. . .a desire to seek, to question, and to contemplate is latent in all intelligent beings. The seeds of a vigorous mental life may be sown at school and bear fruit later when the adult mind is enlarging its sphere of action and is trying to find even a partial answer to the riddles of the universe. (Sumner 1947:12)

A further reason for fostering children's interrogative skills is that, by posing questions, pupils are shaping and exposing their thoughts and hence opportunities will be provided for teachers to have some insight into children's thinking and conceptual understanding. Questions asked by children can lead teachers towards making appropriate assessments of children's understanding or alternatively their misconceptions.

Hence, it is desirable that children *do* ask questions for a variety of reasons, not least so that they acquire a way of addressing new problems, new situations with confidence and competence.

Having recognised the benefits of children's questioning to their own learning, we will proceed with a consideration of the role of teachers' questioning and the way in which this may or may not contribute to the learning process.

Teachers' Questions

A considerable amount of literature is available from research on teachers' questioning, investigating aspects such as the types and frequency of questions asked of pupils, pupils' responses and wait time given for responding. Some of these research findings have direct relevance for our discussions and will be pursued where appropriate, albeit briefly.

It has been advocated that one of the purposes of teachers' questioning behaviour is to provoke thinking in pupils, stimulating learners to focus on specific issues under discussion, and hence creating opportunities for understanding. Indeed, questioning by teachers is seen by many as an essential skill, crucial to their success as practitioners in the classroom. Cohen & Manion (1989:140) in their guide to teaching practice point out that questioning functions 'to elicit information. . .to probe the extent of children's prior learning. . .to revise earlier learning. . .or consolidate recent teaching and learning'. Pollard & Tann (1987:139) comment that questions 'can be seen as a vital tool for teaching and learning', Harlen (1992:109) argues that 'asking questions is an important dimension of teaching' and Kyriacou (1991:37) claims that 'questioning skills are central to the repertoire of effective teaching'. Numerous writers, therefore, are in agreement that teachers' ability to question is an essential part of their practice.

In contrast, however, Dillon (1982) casts doubt on the assumptions made about teachers' questions. He claims that no firm evidence supports the view that questioning stimulates pupils' thought and discussion. Dillon maintains that the emphasis on questions in education may be characterised as 'presumptive practice'—describing the idea that questions stimulate thought and discussion as 'received opinion, conventional wisdom or presumptive knowledge'.

Dillon (1982) finds that, in enterprises other than education, questions are not advocated as a way of stimulating thought and discussion. On the contrary, in various fields of practice 'questions are a very good means to keep people from talking'. Dillon cites pollsters and cross examiners, whose techniques need to reduce clients' thinking and talking, who use questions as a way of initiating only the briefest of replies. Again, in contrast to the work of teachers, professional work involving counselling and personnel interviewing, which requires clients' participation in discussion and inquiry, avoids the deployment of questions. The outcomes are similar to those espoused by teachers—thought, exploration of ideas, expression, but whereas one group avoids the use of questions, teachers actively seek to apply an interrogative strategy.

McNamara (1981), also on a cautionary note, alerts us to the possibility that questioning as a strategy described in method texts focuses on 'the skill *per se* without any discussion on the *validity* of that skill'. He queries the wisdom of accepting the commonly held view reflecting the advantages of questioning and stresses the need for teachers to think more critically about their questioning technique. This is particularly so in view of research evidence cited by McNamara claiming that questions may inhibit rather than foster learning.

If we accept Dillon's view that teacher's questions do not necessarily stimulate thought and discussion, then it is essential to consider alternative ways in which teachers can evoke these positive responses from their pupils. Dillon (1988a), who has commented extensively on the place of questioning in education, recommends that, in order to serve a pedagogical purpose in the classroom, teachers' questions still have a role provided they are of a type which will engender genuine discussion rather than the 'guess what the teacher's thinking' type. Further recommendations proposed by Dillon (1988a) are that teachers should ask fewer questions, pupils should be given more time for response and, importantly, pupils' own questions ought to be encouraged.

To some extent, children learn their questioning skills from their teachers. Hence, in pursuing our aim to encourage children to raise questions it is necessary for teachers to present a model which can be emulated by the pupils. This is particularly the case where children are in the early stages of learning how to pose a question in a way which can be followed by an investigation. It is suggested therefore, that one of the primary aims in science should be to encourage children's questions. Further, this should be extended by helping children to refine their questioning technique in order that they acquire the skill of asking investigable questions. An added advantage of this strategy is that ensuing investigations will have a personal relevance for the children—a criterion which is sometimes overlooked in planning.

Therefore, despite recognising the limiting effect of certain questions used by teachers, it is suggested that if the distinctions between different types of questions are identified and acted upon, in addition to adopting a more sensitive awareness of their timing, pacing and context, then teachers' questions *do* have a part to play in supporting children's learning. Willig (1990) reminds us that skillful questioning lies at the heart of the cognitive conflict strategy whereby children are forced to reflect on their ideas and their reasons for holding these ideas.

The State of the Art

In sharp contrast to the numbers of studies carried out on teachers' questioning, few studies have been put into effect to investigate children's questioning, particularly in primary science. Our observations, therefore, are necessarily restricted by that which is available.

Dillon (1988b) points to the dearth of research related to students' questions, not because there is lack of interest in the issue, but claiming that 'investigators can scarcely find any student questions'. In his study Dillon is forced to conclude:

> . . .children *qua* students do not ask questions. They may be raising questions in their own mind during the class hour. They may be questioning as they read and study their texts. They may be asking questions of their friends and family and of adults in other roles or contexts. But they do not ask questions aloud in the classroom. Dillon (1988b:200)

The findings of Tizard & Hughes (1984) also lend support to this view, pointing to the contrast between the wealth of questions pre-school children ask of their mothers, yet quickly adapting to the classroom situation where their role changes from being the ones who ask to ones who answer questions.

A study based at Queen's University, Belfast is informative for science practitioners seeking to address the issue of pupil questioning. Jarman (1991) in a project working with primary and secondary teachers of science asked the teachers to keep a diary in which they recorded the questions raised by pupils during a four week period. The group found that:

> . . .children *did* ask science-related questions, though fewer than we might have forecast—certainly fewer than we would wish. (Jarman, 1991:7)

The questions children asked were categorised into those in which children sought:

> direction,
> clarification,
> reassurance, and
> information.

Within the fourth category, in which children sought information, Jarman and her team found that a small but significant sub-set of the questions were those which could be followed up with practical investigations in the classroom. In addition, it was found that children also asked questions which had potential for pursuing further but as yet were not in a form which would lead directly into investigations. In view of this Jarman (1991:9) stresses:

> . . .the role of the teacher. . .in interacting with the pupil's action or statement is crucial in recognising that a question has been posed and in helping the pupil to reformulate it.

Before leaving this section in which we have briefly discussed pupil questions, it is pertinent to refer to evidence from two projects whose findings on teacher questioning are also significant for our study. Research by Galton *et al.* (1980) found that time spent in questioning by teachers in junior classrooms amounted to 12% of teacher activity. The type of questions most closely related to promoting thought and imagination represented only 5% of the total questioning i.e. a very small proportion of the teachers' total activity.

Kamara (1983) refers to the findings of Gyang (1975), who analysed questions asked by teachers in primary science lessons, and notes that within the classification of knowledge questions, reasoning questions and activity type questions, most teachers' questions fell into the first category i.e. questions requiring recall.

It would seem therefore, that, not only are pupils' questions under-represented in the classroom but, in addition, the type of teachers' questions which stimulate children's thinking are also of a low proportion.

Why is Pupil Questioning in Primary Science a Problematic Area?

Having established the benefits of children's questions to their own learning and understanding and noting the apparent lack of meaningful questioning that does occur in primary science, then it would seem pertinent to ask why this aspect of science activity is so poorly represented. Clearly, without the benefit of substantial research evidence in this area any reasons offered stand simply as conjecture which, nevertheless, may add to the current thinking on the issue.

We will reflect initially on teachers' questions and speculate as to the under-representation of questions which enhance children's thinking skills.

Much current discussion on primary science is focused on teachers' apparent lack of background knowledge in science and hence the assumption that difficulties will arise in teaching this curricular area. Bennett & Carre (1991:14), in a study where the impact of subject knowledge on teaching performance was observed conclude that:

> . . .subject knowledge is a vital ingredient in high quality teaching and pupil learning. Teachers cannot teach well what they do not know.

One interpretation of this comment could be that if higher order teacher questions are one of the hallmarks of 'high quality teaching' then those who do not have the subject knowledge are restricted by the background from which they can draw higher order questions. Does this therefore, call into question the ability of 'non-specialists' to engage in 'high quality teaching'?

This point is taken up to some extent by Prawat (1989:322) who claims that teachers need to know where they are heading:

> The teacher can be an effective guide only if he or she has a good sense of direction; not having a sure grasp of the cognitive territory one is to traverse puts teachers in the position of the 'blind leading the blind'.

Such observations suggest that lack of background knowledge restricts teachers to asking questions with which they feel comfortable and such questions may not therefore be of the kind to extend children's thinking.

Prawat (1989) cites the work of Hashweh (1985) who found that teachers who were 'knowledgeable about the topics they were teaching. . .were able to generate a better set of questions to assess student understanding'.

Not all writers, however, are in agreement with the over-riding need for subject knowledge. Morrison (1989) warns of placing too much emphasis on content, which, in reality, needs to be balanced by considerations of how children learn, teaching strategies and communcation.

Perhaps, therefore, it is not merely a case of having/not having the subject knowledge but more a matter of knowing/not knowing how to teach effectively. In other words, it needs also to be stated that those having a depth of subject knowledge must also acquire and apply the skills of effective questioning in order to be engaged in 'high quality teaching'.

Hence, one of the possible reasons for teachers' reluctance to ask higher order questions of their pupils is a lack of confidence in their own understanding of the subject area and therefore an avoidance of issues that may prove to be problematic. A further possibility is that teachers' pedagogical stance may be in conflict with a methodology which embraces open discussion and such teachers will avoid strategies which are contrary to their usual practice. This might particularly be the case for science if teachers were subjected to a didactic, knowledge-based approach during their own experiences as pupils.

Biddulph *et al.* (1986:84) lend support to this view suggesting that teachers' views of primary science may be such that questioning does not feature as a pedagogical requirement. These teachers may hold the view that 'teaching science means transmitting a predetermined body of scientific knowledge to children'. The need for higher order questioning therefore does not arise.

Research by Delamont (1983) indicates that one use of questioning by teachers is as a means of control in the classroom—control of pupils' behaviour and control of the direction of discussion. We would suggest that where teachers' uncertainties do not allow them to relax their control then, inevitably, questioning strategies which could involve divergence and digression will be avoided.

The reasons cited above as possible explanations for the apparent paucity of higher order questioning may also be proffered as grounds for lack of pupils' questions. For example, teachers who feel unsure of their own knowledge base may tactically avoid or repress pupils' questions in science. Similarly, teachers who interpret science teaching as transmission of facts and those for whom tight control is a feature of teaching, are also unlikely to invite pupils' questions.

Given the recent curriculum initiatives which primary teachers in England and Wales have had to adopt, it would not be surprising to find that teachers perceive an open style of questioning as a threat to their tight time schedules. The national curriculum science initiative can be viewed perhaps from two perspectives; on the one hand it has, at least, ensured that science experiences are on the agenda for every primary pupil, but on the other hand, the specification of so wide a content area may suppress opportunities to implement the curriculum in ways which are most effective for pupils' learning for fear of taking too much time.

Should the reasons for lack of questioning in primary science, discussed above as conjecture, be grounded in evidence then clearly this would have far-reaching implications for training. Both initial and inservice training for teachers have a vital role in raising awareness of the issues, equipping teachers to carry out their work effectively, and supporting them in reflecting on their classroom practice.

Creating Opportunities for Questions

In the previous section we focused on the apparent low priority accorded to children's questions and productive teachers' questions in science activity. Science teaching and learning deprived of motivating questions, is an impoverished state of affairs which does not take advantage of children's potential enthusiasm for genuine involvement. We will thus consider ways of addressing this deficiency with reference to the work of those influential in this field.

White (1977) bemoans the fact that students spend much of their time in formal education learning how to answer questions not how to ask questions. He suggests that students need to be equipped with the ability to formulate questions and this is a skill which needs to be taught rather than left to chance. Numerous writers echo this plea (e.g. Dillon, 1988a; Jelly, 1985; Harlen, 1992; Morgan & Saxton, 1991).

These writers believe that it is not enough merely to provide opportunities for children's questions, but a strategy needs to be operated which takes a more positive stance. Jarman (1991) and Jelly (1985) are but two who advocate that teachers should provide models for pupils in the way that they create and pose questions. Teachers whose scientific background is limited may indeed be at an advantage here, in that the questions they ask may be genuinely 'a need to know'—hence they would investigate alongside pupils in a truly collaborative way.

Dillon (1988a) and Jarman (1991) point to the need for praise and positive reinforcement in accepting pupils' questions. Dillon (1988a: 28) sees this as a 'particularly crucial move when the first students begin to ask and when the student first begins to ask'. The importance of creating a classroom climate where pupils feel comfortable and confident asking questions is cited by several writers as an important factor in developing children's questioning. Such a climate is particularly significant for conveying the message that science is an area in which inquiry is a natural component and questions need constantly to be raised.

Morgan & Saxton (1991) state that training students to question should be composed of two stages—informing students of the reasons for questioning—the how and why of asking questions, followed, of necessity, with adequate opportunities to practice. Jarman (1991) and Jelly (1985) list a variety of strategies which can be implemented in the primary classroom to encourage children to practise such questioning techniques.

The importance of the provision of stimuli in order to provide a springboard for children's questions is identified by Harlen (1985) and Jelly (1985) as a further way of creating opportunities for children's questions. Finally, Biddulph *et al.* (1986) and Jarman (1991) point to the relevance of questioning during the evaluation of a science activity. Questioning at this stage can extend children's thinking, for example: 'How could we have improved that activity?' and also help the teacher to reflect on practice, for example: 'Did my questions help to extend children's thinking?'

However, making increased provision for children's questions goes only part of the way to meeting the potential of questioning behaviour. What needs to be considered in tandem, is the effectiveness of teacher questioning. Elstgeest (1985) offers guidelines to teachers as to the type of questions which are 'productive' for primary science. In addition, within the productive category, Elstgeest identifies a hierarchy of question types which can be posed according to children's experience of a particular topic.

Other writers (e.g. Dillon, 1988a; Morgan & Saxton, 1991), although not referring specifically to primary science, nevertheless offer guidance as to how teachers, by evaluation and reflection on their own practice can learn to ask more effective questions and to use questioning in a more effective way.

Biddulph & Osborne (1984) have addressed directly the need to provide support for teachers of primary science, by producing material whose aim is partly directed at advising teachers on how to generate and use children's questions in science. Clearly such guidance and support is particularly valuable in an aspect of primary teaching which is proving to be problematic.

Conclusion

It has been argued in this paper that the requirement to encourage children's questioning in science is educationally acceptable and indeed beneficial. Furthermore, it is believed that teachers' questions, with certain provisos, can contribute positively to children's learning in science.

The issues most evident in the study are first, that much contemporary classroom practice does not embrace the opportunity for effective questioning and, perhaps of greater significance, many teachers of primary science need substantial guidance in developing their own and their pupils' questioning strategies.

In conclusion, it is suggested that provision needs to be made available for primary teachers to reflect on their current practice. Where such a high proportion of practitioners' time is taken up with receiving, interpreting and implementing new intiatives, one has to ask whether indeed this necessarily results in improved practice. It is suggested that the latter will arise only when teachers are given time to discuss, review and evaluate what they do, so that alternative and improved ideas, informed by reflection, can be tried out. Researchers outside the classroom may propose relevant and significant changes but, unless teachers themselves have opportunities to be involved in reflective practice then it is tempting to suggest that the status quo will prevail. This is perhaps best summarised by Stenhouse (1975:143) whose comment could be applicable to any curricular area:

> It is not enough that teachers' work should be studied; they need to study it for themselves.

References

Bennett, N. and Carre, C. (1991) No substitutes for a base of knowledge. *Times Educational Supplement* 8.11.91. p.14.

Biddulph,F. and Osborne, R. (eds) (1984) *Making Sense of our World: An Interactive Teaching Approach.* Hamilton, New Zealand: SERU, University of Waikoto.

Biddulph, F., Symington, D. and Osborne, R. (1986) The place of children's questions in primary science education. *Research in Science and Technological Education* 4 (1), 77–88.

Cohen, L. and Manion, L. (1989) *A Guide to Teaching Practice* (3rd edn). London: Routledge.

Delamont, S. (1983) *Interaction in the Classroom* (2nd edn). London and New York: Methuen.

Department of Education and Science (1991) *Science in the National Curriculum.* London: HMSO.

Dillon, J. T. (1982) The effect of questions in education and other enterprises. *Journal of Curriculum Studies* 14 (2), 127–52.

——(1988a) *Questioning and Teaching. A Manual of Practice.* London and Sydney: Croom Helm.

——(1988b) The remedial status of student questioning. *Journal of Curriculum Studies* 20 (3), 197–210.

Elstgeest, J. (1985) The right question at the right time. In W. Harlen (ed.) *Primary Science—Taking the Plunge.* London: Heinemann.

Galton, M. J., Simon, B. and Croll, P. (1980) *Inside the Primary Classroom.* London: Routledge.

Gyang, M. (1975) Analysis of questions asked during classroom interactions in Science lessons. Unpublished thesis for certificate in Science Education, Njala University College.

Harlen, W. (1985) *Teaching and Learning Primary Science.* London: Harper and Row.

——(1992) *The Teaching of Science.* London: David Fulton Publishers.

Jarman, R. (1991) *Questioning. An Important Science Skill.* Queen's University Belfast:ASE/ICI/NIESU.

Jelly, S. (1985) Helping children raise questions—and answering them. In W. Harlen (ed.) *Primary Science—Taking the Plunge.* London: Heinemann.

Kamara, A. (1983) The training of science teacher educators in Africa. In W. Harlen (ed.) *New Trends in Primary School Science Education Vol 1.* Paris: UNESCO.

Kyriacou, C. (1991) *Essential Teaching Skills.* Oxford: Blackwell Education.

McNamara, D. R. (1981) Teaching skill: The question of questioning. *Educational Research* 23 (2), 104–9.

Morgan, N. and Saxton, J. (1991) *Teaching Questioning and Learning.* London and New York: Routledge.

Morrison, K. (1989) Training teachers for primary schools: The question of subject study. *Journal of Education for Teaching* 15 (2), 97–111.

National Curriculum Council (1991) *Science in the National Curriculum. Consultation Report.* York: NCC.

Pollard, A. and Tann, S. (1987) *Reflective Teaching in the Primary School.* London: Cassell.

Prawat, R. S. (1989) Teaching for understanding: Three key attributes. *Teaching and Teacher Education* 5 (4), 315–28.

Stenhouse, L. (1975) *An Introduction to Curriculum Research and Development.* Oxford: Heinemann.

Sumner, W. L. (1947) *The Teaching of Science* (3rd edn). Oxford: Blackwell.

Tizard, B. and Hughes, M. (1984) *Young Children Learning: Talking and Thinking at Home and at School.* London: Fontana.

United Nations Educational, Scientific and Cultural Organization (1980) *UNESCO Handbook for Science Teachers.* Paris, UNESCO/London: Heinemann.

White, R. T. (1977) An overlooked objective. *Australian Science Teachers Journal* 23 (2), 124–5.

Willig, C. J. (1990) *Children's Concepts and the Primary Curriculum.* London: Paul Chapman Publishing.

CO-ORDINATING SCIENCE IN PRIMARY SCHOOLS: A ROLE MODEL?

Derek Bell

Liverpool Institute of Higher Education, Stand Park Road, P.O. Box 6, Liverpool L16 9JD, UK

Abstract The number of primary schools with a teacher responsible for co-ordinating and developing science in the curriculum (a science co-ordinator) has increased dramatically since 1978. However co-ordinators are not having the impact they should on the quality of science in the schools. This in part is due to the restricted view of the role that is held by headteachers, co-ordinators and classteachers. The nature of the science co-ordinator's role is considered and the way in which it is perceived discussed. A model for appraising the role of the science co-ordinator more thoroughly is proposed and ways of applying the model are considered.

Introduction

Co-ordinating any curriculum area in a primary school is a very complex process. It involves all members of staff but is the particular responsibility of individual curriculum co-ordinators. The role of the curriculum co-ordinator has developed greatly in recent years due to the general trends and changes that have taken place in primary education. The role of the science co-ordinator is no exception. Indeed in the UK it is perhaps the most outstanding example as a result of the dramatic change in the status of science through its inclusion as a core subject, alongside mathematics and English, in the National Curriculum as laid down in *The Education Reform Act 1988* (Laws and Statutes, 1988).

The increased recognition that science should be a major component of the primary school curriculum, has been reflected in the number of primary schools having teachers with responsibility for co-ordinating and developing science within the curriculum. The number rose dramatically from 17% in 1978 (DES, 1978) to approximately 70% in 1984 (DES, 1988a) and there has been a further increase since (DES, 1989). However, despite this recognition, the report by HMI, *Aspects of Primary Education: The Teaching and Learning of Science* (DES, 1989), indicates that the effectiveness of the science teaching observed in primary schools is still below the standards required. It also reports (paragraph 29) that the teachers with responsibility for science are not having the impact they should, with only one in five succeeding in influencing the work throughout the whole school. The reasons for this lack of influence are not always clear but as stated by HMI in *Primary Education in England* (DES, 1978):

. . . where a teacher with a special responsibility was able to exercise it through the planning and supervision of a programme of work, this was effective in raising the standards of work and the levels of expectation of what children were capable of doing. (paragraph 4.6).

This would suggest that there is a strong argument for improving the performance of co-ordinators as part of an overall strategy to improve the teaching of science in primary schools. However it would seem that a major obstacle to any improvement is the way in which the role is perceived by headteachers, co-ordinators and their colleagues (Bell, 1990). This paper will:

- review the nature of the science co-ordinator's role;
- examine how the role is perceived by headteachers, co-ordinators and class teachers;
- suggest a possible model which can be used to appraise the role more thoroughly.

The Nature of the Science Co-ordinator's Role

Origins of the science co-ordinator's role

Descriptions of the role of teachers with specific responsibility for co-ordinating and developing science in primary schools are few. Taylor (1986: Chapters 1 and 2) discusses the issues related to the overall problem of the expertise and responsibility of the primary school teacher and provides a summary of the arguments which have been put forward since the beginning of the century. One of the key issues running through the debate is the role of teachers who have a particular responsibility for some aspect of the curriculum. In essence it is argued that the discharge of such responsibility should involve the teacher acting as either a 'specialist teacher' or a 'curriculum co-ordinator'. The former would see the specialist teaching his/her subject to different classes throughout the school. The latter, on the other hand, would see the teacher concerned acting as a consultant to other staff and taking some responsibility for the development of the curriculum area throughout the school. The 'curriculum co-ordinator' role does not involve the post holder in taking other classes for that particular curriculum area on a regular basis. Although it can be argued that subject specialists have a place in primary schools, it is the 'curriculum co-ordinator' role which is the concern of this paper.

The detailed and extensive study of primary education in England and Wales carried out by Blyth (1965) in fact identifies a variety of roles for the primary school teacher but makes no reference to teachers taking responsibility for a curriculum area across the school. Thus it would seem that up to the mid 1960s few, if any, teachers could be regarded as curriculum leaders or co-ordinators. Since then, however, the situation has changed markedly. Several major reports (CACE, 1967; DES, 1975; DES, 1982a), Government documents (e.g. DES, 1983a; DES, 1985a) and HMI (e.g. DES, 1982b; DES, 1983b) have all reinforced the notion that primary school teachers should have a responsibility for some aspect of the curriculum. In 1988 it was written into the terms and conditions of service (DES, 1988b) that teachers, in addition to their role as a classteacher, should have a responsibility for some aspect of the curriculum.

At the end of the 1970s, although in a minority of primary schools science was established and of good quality, the majority of schools did little or no science.

However HMI were pointing a way forward to improve the situation by encouraging schools to use the expertise of a science co-ordinator. This was also proposed in many subsequent publications from the Association for Science Education (ASE, 1981) and HMI (e.g. DES, 1982b, 1983b, and 1985c). Perhaps the most significant document to be published in relation to science education at this time was *Science 5–16: A Statement of Policy* (DES, 1985b). This, for the first time, set out a clear, generally agreed statement of the need to have a coherent policy for science as an essential part of the education of all children. Indeed this policy statement singled out science as a curriculum area which required special consideration.

For primary schools, *Science 5–16* lists five features that are regarded as crucial if science is to develop. These include the statement:

> . . . the school needs at its disposal at least one teacher with the capacity, knowledge and insight to make science education for primary pupils a reality; in the case of small schools such consultant teachers may need to offer advice to more than one school; (paragraph 20)

Then the document goes on to outline the role of such teachers:

> They can act as science consultants or experts in the primary school, stimulate science teaching throughout the school and provide help and support for their colleagues. This support may take the form of assistance with the preparation of programmes of work, individual lessons or materials, and it may involve taking on part of the teaching of some classes, for example with older pupils. Encouragement and support from the headteacher and from the LEA are also essential. (paragraph 21)

The commitment of DES to support the development of science in primary schools was demonstrated when LEAs were invited to apply for money from the Educational Support Grant scheme (DES, 1984a) in 1984 and again in 1985. The Association for Science Education were commissioned to carry out an evaluation of the schemes run under these arrangements. The *Initiatives in Primary Science: An Evaluation (IPSE) Report* (ASE, 1988a) that was subsequently published included the recommendation that: 'All but the smallest schools should designate a teacher as the "co-ordinator for science"' (page 50).

The role of the science co-ordinator

Detailed studies of the role of curriculum co-ordinators are few. The main ones are Rodger (1983), Campbell (1985), and Stow & Foxman (1988). In addition Taylor (1986) refers extensively to the studies of the Primary Schools Research and Development Group (1983) at the University of Birmingham, and Morrison (1986) uses his own previously unpublished findings in discussing the role of curriculum co-ordinator. Although there have been no major studies specifically referring to science co-ordinators, it is reasonable to assume that the main findings of these other studies apply to science.

Morrison (1986) compares the reports and notes that all of them indicate curriculum co-ordinators should have expertise in curriculum development, management skills, innovation of new ideas, current thinking in the curriculum area, dissemination and training, and resourcing. He further analysed the key areas of responsibility noted in the studies by Rodger (1983) and Campbell (1985) against his own findings. Correlation was very high between the ten key tasks which, in rank order, were shown to be:

Communication with the headteacher.
Exercising curriculum leadership.
Communication with staff.
Organisation of resources.
Establishing and maintaining continuity throughout the school.
Organising in-service work.
Liaison between head and staff.
Establishing record systems.
Motivating staff.
Curriculum development.

Thus there seems to be at least some agreement as to the role of a curriculum co-ordinator and that it requires the same basic qualities of the incumbent regardless of the curriculum area concerned. However there are some specific demands made on individual co-ordinators which are closely related to the nature of their particular curriculum area. Science perhaps makes greater demands than other areas.

Various sources have defined the role to be undertaken by the science co-ordinator. Many of them provide a list of activities, such as that suggested by Harlen (1985:214). LEAs have also indicated in their policy statements how the co-ordinator might carry out their role. An evaluation of LEA guidelines (Anon 1987) from 18 authorities produced the following list:

 (1) Good working relationships with colleagues.
 (2) Working alongside members of staff to give practical classroom help.
 (3) Keeping up to date with new ideas.
 (4) Attending courses. Sending back information to staff.
 (5) Encouragement of staff to attend courses.
 (6) Liaison with Head to develop with staff a school policy.
 (7) Development of schemes of work which are understood by all staff.
 (8) Determine resources needed for science.
 (9) Liaison with outside agencies.
(10) Organising in-service sessions at school for staff.
(11) Able to act in an advisory capacity or consultant role to help colleagues.
(12) Make staff aware of the health and safety aspect in the teaching of science.
(13) Monitor the operation of the school science policy and recommend appropriate changes.
(14) Assist staff to devise programmes of work to cater for children with special needs.

The very helpful booklet *A Post of Responsibility in Science* produced by ASE (ASE, 1981) addresses many of these points emphasising in particular the need to develop working relationships with headteacher, other staff, children, and advisory staff. Raper & Stringer (1987:70) suggest that the co-ordinator should be a 'model of competence', enthusiastic, 'an authority', and possess 'advisory skills'. The ASE booklet (ASE, 1981) identifies the need for a teacher who has:

- a good 'understanding of children's development and a genuine concern for colleagues'.
- 'qualities of leadership, tact and enthusiasm for science learning'.

- an informed grasp of the curriculum area.

The booklet stresses that: 'the attitude of the holder of the scale post is of far more importance than traditional knowledge and qualifications'. It goes on to say:

> Teaching methods and subject matter constantly change. Somebody willing to admit, at times, to not knowing, but alert to the need for finding out and for keeping up to date, is likely to be active in seeking the help that science advisers and advisory teachers can provide. Such a person will also use in-service training, meetings at Teachers' Centres and relevant books and articles, and encourage other members of staff to take advantage of the same opportunities. (p. 50)

However there is a danger that many of the references to the role of science co-ordinator include statements that are generally applicable to all co-ordinators, with the result that the specific requirements of science are not catered for. Raper & Stringer (1987:68–9) highlight some of these points as they recognise the similarities between P.E. and science. Thus they point out that science is:

- essentially practical and activity based
- utilises equipment and so involves financial outlay, maintenance, running costs, replacement and repair.
- can lack progression by using the same activities with young children and essentially repeating them with older pupils but failing to draw out the extended concepts and ideas.
- ephemeral in that no matter how detailed a record of the activities it can never show exactly what happened or how the child reacted.

In addition there are difficulties which might arise because of:

- the need for the development of personal scientific knowledge,
- the need for safety procedures specific to science,
- the extent and variety of resources and equipment required,
- the fact that a high proportion of the staff lack confidence and training in science work.

These requirements as well as the more general ones have increased in importance with the establishment of The National Curriculum in which science is a core subject. Now the task of ensuring that every child is taught science effectively is essential and not just desirable. Therefore the establishment of a science co-ordinator in every school is only a start, because the teacher designated is required to ensure the delivery of an effective science curriculum in the school.

Perceptions and Performance of the Role of the Science Co-ordinator

Although there is general agreement in the literature of what the role of a science co-ordinator should involve, the degree of influence that co-ordinators have in schools depends on many factors. In particular there needs to be:

- a shared perception of the role between the headteacher, the co-ordinator and the classteachers;
- a match between the demands made in the school situation and the performance of the co-ordinator.

These two issues were addressed in a recent study of science co-ordinators in three LEAs in the north-west of England (Bell, 1990). Headteachers, co-ordinators and classteachers were asked to rate various tasks in terms of their importance in the role of the science co-ordinator and the effectiveness with which the tasks were carried out.

Perceptions of the science co-ordinator's role

The list of tasks covered the range of activities one might expect a co-ordinator to carry out. Very few items were added by respondents suggesting, at least superficially, that the tasks provided an acceptable description for the role of a science co-ordinator. The most significant finding, statistically at least, was the overall agreement on the level of importance of the various tasks shown by the groups of teachers involved in the study. Of particular note was the fact that in general headteachers, co-ordinators and classteachers seemed to have a similar perception of the priorities for a co-ordinator.

The tasks can be placed into four groups according to their perceived importance as a result of the responses made. These groups are summarised in Table 1. The items considered most important (Group A) probably reflect not only the actual needs of the school but also some of the pressures from outside, notably

Table 1 Comparison of ranking groups (A,B,C and D) for importance and effectiveness ratings for tasks considered part of the co-ordinator's role

Ranking Group	Importance Rating	Effective Rating
A	Communication with staff Develop a school policy for science Organisation of resources	Communication with headteacher Organisation of resources
B	Keep up-to-date Provision of resources Communication with headteacher Suggest approaches and starting points	Communication with staff Develop a school policy for science Provision of resources Suggest approaches and starting points Keep up-to-date
C	Evaluation of science in curriculum Provide opportunities to improve staff expertise and knowledge in science Monitoring of pupils' progress in science Liaison with LEA	Evaluation of science in curriculum Provide opportunities to improve staff expertise and knowledge in science Liaison with LEA Liaison with outside bodies
D	Work alongside colleagues in their own classroom Liaison with other schools Liaison with outside bodies	Monitoring of pupils' progress in science Liaison with other schools Work alongside colleagues in their own classroom

the requirement by the LEAs for each school to have a policy document for each area of the curriculum. The organisation of resources was seen as a crucial task which was frequently commented on as a necessity if teachers were going to include science as part of the curriculum.

Tasks in Group B would appear to meet the initial needs of the classteachers (provision of resources and suggestion of approaches and starting points) and the expectation that the co-ordinator should know about primary science.

Liaison work especially that with other schools and outside bodies is given a low priority probably because they do not seem to have a direct effect on the teaching of science in the school. Indeed approximately 25% of the respondents regarded these tasks of minor importance, unimportant or just irrelevant.

Evaluation of science in the curriculum and the monitoring of children's work were considered important but not essential to the job. Some of the replies indicated that both of these were irrelevant or at least not part of the co-ordinator's role. This would seem to be a very narrow view especially in the context of the demands which are being made on schools and teachers as a result of the Education Reform Act of which the National Curriculum is only a part.

The remaining two items ('working alongside colleagues in their own classrooms', and 'providing opportunities for in-service work') are closely related and involve the co-ordinator in improving the expertise of the other teachers in the school. Given the need for this type of activity because of the lack of confidence and scientific knowledge of the vast majority of primary school teachers, it might have been expected that these would have been given a greater importance rating. However many of the comments that were made by various respondents suggested that these tasks were the responsibility of others such as advisory teachers. Whilst co-ordinators were willing to perform these tasks informally in a reactive manner, few felt they needed to initiate activities specifically to improve teachers' expertise in science.

Performance of the science co-ordinator's role

Perceptions of how effectively the co-ordinators carried out their role showed a similar level of agreement between the headteachers, co-ordinators and the classteachers. It was again possible to group the tasks, this time according to how effectively they were carried out (see Table 1). Without exception communication with the headteacher and organisation of resources were the tasks rated as being carried out most effectively, thus making up Group A (see Table 1). The tasks making up Group B were all rated very close together at better than average. It would seem that the tasks in groups A and B were in general seen to be carried out satisfactorily.

However the tasks in Groups C and D for effectiveness were below average and needed to be improved. Significantly these include those tasks which relate to evaluation, monitoring, liaison, and in-service. As with the ratings of importance for these tasks, the levels of effectiveness give rise for concern given the demands of the National Curriculum.

Comparing perceptions and performance

Comparison of the importance and effectiveness ratings of the tasks carried out by co-ordinators shows close statistical agreement (see Table 2), so that on the

Table 2 Comparison of importance rankings with effectiveness rankings for tasks considered part of the co-ordinator's role

Analysis grouping	Spearman rank co-efficient	p-value
All responses	0.86	0.0001
Headteacher responses	0.90	0.0001
Co-ordinator responses	0.88	0.0001
Classteacher responses	0.89	0.0001
Infant school responses	0.86	0.0001
Junior school responses	0.80	0.0006
Primary school responses	0.87	0.0001
LEA 1 responses	0.90	0.0001
LEA 2 responses	0.91	0.0001
LEA 3 responses	0.80	0.0005

whole the items considered most important were carried out the most effectively. However, as Table 1 shows, the actual positions of the tasks in each list did vary.

The differences in the position of tasks in the two lists implies that some activities, whilst being highly desirable prove more difficult to carry out in practice and conversely those which are considered less important are easier to put into practice. Of particular note might be 'communication with other members of staff' and 'development of a policy for science'. These were rated the two most important tasks, but they were not carried out as effectively as their importance rating might suggest. The major reason given by respondents was that of time, in particular the time to get together with the whole staff to discuss science. In contrast 'communication with the headteacher' was considered the most effectively carried out task but its importance was much lower. Presumably this was because, in practice, it is easier to meet with one person and because many of the decisions to be taken by the co-ordinator required some comment by the headteacher before action could be taken.

The findings of this study (Bell, 1990) show a remarkable correlation with those given in the IPSE report (ASE, 1988b), and reproduced here as Table 3. Tasks

Table 3 Co-ordinators' perceptions of their role (from IPSE report 1988)(source ASE, 1988b)

Task	Number (n=84)	Percentage
Advise and encourage staff	84	100
Management of resources	71	85
Establish and implement policy	29	35
Keep up to date with national and local initiatives	24	28
Plan development of science in school	14	16
Run INSET/workshops for staff	13	15
Assist staff with termly planning	11	12
Encourage uptake of INSET	11	12
Work alongside colleagues	7	8

related to communication with the other staff including the headteacher and the management of resources are given a high priority. Development of 'whole school' approaches through such tasks as establishing a policy and continuity within the school tend to follow while working with colleagues and carrying out in-service work tend to be towards the bottom of the priority list. Similarly liaison activities with other schools and/or outside agencies including LEAs are not considered to be top priorities.

A Model for the Science Curriculum Co-ordinator

The results of the study by Bell (1990) show that, in general, headteachers, co-ordinators and classteachers have a common perception of the role of the science co-ordinator. However when this perception is compared with the list of tasks given in the literature it would seem that the view of the co-ordinator's role is somewhat restricted. If the influence of the co-ordinator is to be increased then it would seem that there is a need to help co-ordinators, their headteachers and colleagues to develop a wider view of the responsibilities involved. One way in which this might be achieved is in the development of a model for the role of science curriculum co-ordinator. I therefore propose such a model which can be used to appraise the role of the co-ordinator and its performance more thoroughly.

Although studies into the role of co-ordinators acknowledge that each situation is unique, it is clear that there are some essential features of the role which can be identified. These provide the basis for the proposed model.

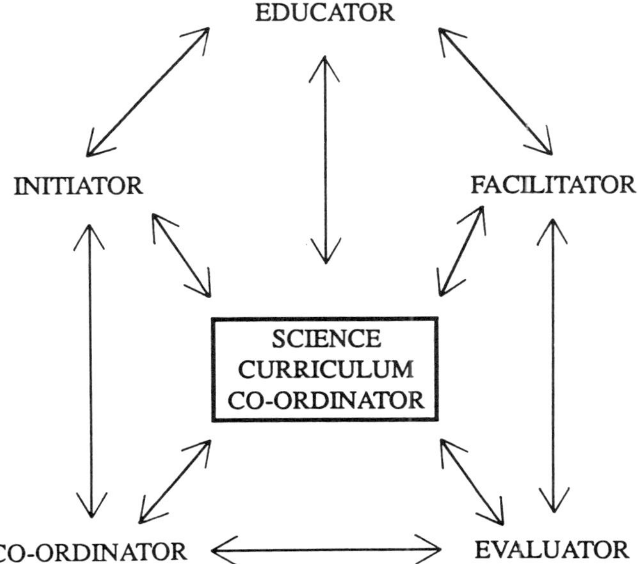

Figure 1 A model for the science curriculum co-ordinator in primary schools.

Figure 1 provides the basic features of the proposed model which shows that the role of co-ordinator consists of five interactive functions, requiring the co-ordinator to:

- *initiate* the development of science teaching and learning in the school;
- *facilitate* the development of science teaching and learning in the school;
- *co-ordinate* the development of science teaching and learning in the school;
- *evaluate* the development of science teaching and learning in the school;
- *educate* others in order to help them develop science teaching and learning in the school.

Whilst specific tasks will often involve aspects of more than one function, it is possible to group the variety of tasks according to the principal function being carried out. For example,

- as an *initiator* a co-ordinator would
    - introduce new ideas;
    - develop a school policy;
    - suggest starting points and ideas for science work;
    - develop a stimulating environment;
- as a *facilitator* a co-ordinator would
    - provide resources;
    - organise storage, accessibility and use of resources;
    - act as a source of information about materials of all types;
    - be available for discussion and consultation;
- as a *co-ordinator* the co-ordinator would
    - compile information on all aspects of science teaching and learning;
    - attempt to ensure continuity of experience for pupils both within the school and beyond;
    - liaise with outside bodies and other schools;
    - oversee matters of safety;
    - arrange staff discussions and training;
- as an *evaluator* the co-ordinator would
    - assess and revise the policy and its practice;
    - identify appropriate forms of assessment of pupils' performance and ability;
    - monitor and evaluate the progress of pupils throughout the school;
    - examine the use and suitability of resources both old and new;
    - examine the degree of integration of science within the whole curriculum;
    - assess schemes of work;
    - improve/develop appropriate methods of record keeping;
    - monitor effectiveness of safety procedures;
    - record staff development and expertise;
- as an *educator* the co-ordinator would
    - help colleagues to understand the nature of scientific activity;
    - help to build up colleagues' confidence in their ability to teach science;
    - provide guidance and suggestions on ways to approach the teaching of science;
    - support staff in their science work;
    - create an awareness of opportunities for staff to develop their own expertise in science;
    - run workshops and similar sessions on aspects of primary science for other members of staff;

- work alongside colleagues in their own classrooms;
- disseminate useful information to keep staff up to date with any major developments in primary science both nationally and locally.

Essential to all of these functions is the ability of the co-ordinator to communicate effectively with the headteacher, colleagues and, when appropriate, school governors, parents, LEA advisers, advisory teachers and others outside the school.

It is important in consideration of this model to recognise that it is dynamic and proactive, not static and reactive. As such the model requires that the role involves the interaction of the five functions, so that any one task is likely to include aspects of more than one function. As it evolves it will take on different emphases at different times according to the expertise of the person carrying out the role, the actual situation being developed and the stage that it has reached.

In order to carry out the role the model also requires that the co-ordinator has appropriate personal qualities which, ideally, would include:

- a strong interest in and enthusiasm for science;
- a positive and enquiring approach to the teaching of science;
- good leadership qualities and interpersonal skills being particularly sensitive to the feeling of other colleagues;
- a good background understanding of scientific processes and concepts;
- the ability to be a good classteacher.

The vital need for these qualities in a co-ordinator is emphasised by HMI in *The Teaching and Learning of Science* (DES, 1989) which says:

> In general, those [co-ordinators] who are most influential are knowledgeable about science and ways of teaching it, are accepted by their colleagues as capable teachers, are possessed of good interpersonal skills, and are well supported by the head. (from paragraph 29)

In summary the proposed model is dynamic and proactive, identifies five functions for a co-ordinator (initiator, facilitator, co-ordinator, evaluator and educator), and requires suitable personal qualities if it is to be put into practice effectively. At this stage the model is tentative and requires further testing in different situations. However it is possible to suggest possible applications of the model and this is done below.

Applying the Model

The model reflects the diversity and wide range of demands that are made on a science co-ordinator in a primary school. Thus it can be used in a number of ways in an effort to increase the effectiveness of co-ordinators in primary schools and hence make a contribution towards improving the quality of the science teaching and learning. The model might be used in:

(a) defining the role of the co-ordinator in a particular school at the time of appointment;
(b) analysing the functions of the role within an existing school situation;
(c) reviewing the co-ordinator's role in the light of changing circumstances both locally and nationally.

A short discussion of each of these uses is provided for exemplification.

Defining the role of the co-ordinator in a particular school at the time of appointment

This is essentially the task of drawing up a job description. Whilst it might be argued that a job description is not needed this is not the conclusion drawn from studies of the co-ordinator's role, for example:

> Job descriptions, jointly arrived at, appear to lead to increased effectiveness on the part of the postholder. (Rodger 1983:137)

However in the study by Bell (1990) less than 20% of the co-ordinators considered themselves to have a job description. In addition, a number of heads indicated that their co-ordinator showed very little appreciation of what the role of co-ordinator might involve. The drawing up of a job description would help the headteacher and the co-ordinator recognise the complexity of the role and identify the initial priorities for the co-ordinator.

The headteacher and co-ordinator could use the model as a common starting point for their discussion and identify which aspects were particularly relevant in their situation. The strengths and weaknesses of science being carried out in the school would have to be considered and priorities identified. The co-ordinator could then select those aspects of the role which most are appropriate to meeting the demands at that time. The important thing is that by using the model all the parties involved would have a clearer picture of the breadth and depth of the role at the outset.

A good classteacher is not automatically a good co-ordinator and a newly appointed co-ordinator is unlikely to have all the necessary qualities to carry out the job effectively. Some of the skills and knowledge will accrue through experience, but by using the model as a guide it should be possible to identify aspects of the role that could be improved by further training. Thus the proposed model could be used as a tool to help in planning the development of the co-ordinator's personal expertise both in science and the managerial aspects of the role.

Analysing the functions of the role within an existing school situation

An important aspect of the management of curriculum development in a school is the evaluation of existing practice as a preliminary to further action and progress. Unfortunately this does not always happen as Clayden (1989) found in her study of developing the science curriculum in 47 schools. Only 13% of the schools had initiated their curriculum development in science with an evaluation of existing practice. However it would seem that such an evaluation is crucial to effective development and that the co-ordinator should play a central role in it. Similarly if the co-ordinator is to influence developments then an analysis of his/her own role should also be undertaken. The proposed model would seem to have a contribution to make to the analysis of the co-ordinator's role and by implication to the planned curriculum development.

An individual co-ordinator could take the model and match his/her performance of the role against the functions given in the model by allocating the tasks which they actually perform to each of the five aspects. This would indicate where he/she was putting his/her effort currently. By matching this against the assessed needs for further development it would be possible for the co-ordinator to decide if there needs to be a change in emphasis of his/her role or to address a particular function more specifically, hence decide priorities.

At a more general level the model can be used to analyse the findings presented in Table 1. If the tasks listed are allocated to the functions of the model, the following distribution is found.

Initiator
  – Develop a school policy for science.
  – Suggest approaches and starting points.
Facilitator
  – Organisation of resources
  – Provision of resources
Co-ordinator
  – Keep up to date
  – Liaison with other schools, LEA, and outside bodies
Evaluator
  – Evaluation of science in the curriculum
  – Monitoring pupils' progress in science
Educator
  – Provide opportunities for staff to improve their own knowledge and expertise in science
  – Work alongside colleagues in their own classroom.

Communication tasks are taken to be essential to all functions.

A comparison of these groupings with the ranking groups for both importance and effectiveness of the tasks (see Table 1) shows that both the Evaluator and Educator functions are found exclusively in Ranking groups C and D. Thus indicating that these two aspects of the co-ordinators' role are not only considered of lower importance they are also done less effectively. Liaison activities which form part of the co-ordinator function are also in the ranking groups C and D (see Table 1). In contrast those tasks which fall into the Initiator and Facilitator functions all occur in ranking groups A and B. Thus it seems that the role of the science co-ordinator is seen as being mainly concerned with the Initiator and Facilitator functions. Other aspects of the role are of lesser importance. This impression matches very closely the observations of HMI (DES, 1989):

> The main tasks they [science co-ordinators] perform, in order of descending frequency, are: acquiring and organising resources (one in three co-ordinators); leading in the preparation of school guidelines; giving informal advice on request; teaching classes other than their own; and working alongside other teachers (one in eight co-ordinators). (paragraph 29)

Reviewing the co-ordinator's role in the light of changing circumstances both locally and nationally

A review of the co-ordinator's role starts with an analysis such as that described in (b) above. This needs to be followed by an identification of any new demands brought about by the changing circumstances which may be related to:

- the individual co-ordinator whose perception of his/her role will change with: experience, increasing confidence in his/her ability to carry out the role, increased appreciation of wider issues related to curriculum development and the management of it;

- school circumstances such as: changes in staff, resources and funding, reorganisation of space and/or the curriculum itself;
- LEA initiatives affecting science both directly and indirectly;
- national directives and recommendations from DES and other bodies.

The introduction of a National Curriculum was a major change for primary schools in the UK. The proposed programmes of study have to be transcribed and incorporated into appropriate schemes of work and there is a much greater emphasis on the assessment of pupils' progress and the need to evaluate the effectiveness of the science teaching and learning within the curriculum. As Taylor (1989) concludes:

> The effective management of the National Curriculum will depend on the effective use of all staff. In this, the curriculum co-ordinator has a vital role to play whether as part of a large staff or as a member of a cluster group of smaller schools.

Thus it is an appropriate time for schools to reconsider the nature of the co-ordinator's role. The proposed model can be used to do this.

The analysis, given above, shows that the role of the co-ordinator is predominately seen as that of initiator and facilitator. For many co-ordinators the emphasis may well need to remain on these functions because of the need to establish a policy and/or scheme of work in addition to the provision and organisation of resources. However the demands of the National Curriculum require an expansion of the role into the other functions.

Educator

The fact that science is a core subject means that all class-teachers are now expected to teach science to their class. Yet the data presented in Table 4 show that the science background of primary teachers is very weak. Thus there is an enormous demand for in-service training to improve the scientific understanding of teachers. It is not possible for all classteachers to go on courses in order to meet this requirement. Much of the work will need to be done in school and the burden will fall on the co-ordinator, hence increasing the emphasis on the educator function. Time will have to be found to allow various in-service activities to take place. In particular time is needed for co-ordinators to work alongside their colleagues, not only providing immediate support but also looking to monitor progress and evaluate future needs.

Table 4 Science background of primary school teachers: percentage who followed a 'science' course as first degree or main subject level in training course

Survey	Percentage
HMI Survey 1979–80 (DES, 1983b) Middle Schools	11
HMI Survey 1981 (DES, 1985c) Combined and Middle Schools	8
APU Survey 1981 All Schools (DES 1983c)	11
APU Survey 1982 All Schools (DES, 1984b)	8
School Survey 1986 (Bell, 1990)	7

Evaluator

The requirements of the National Curriculum for inclusion of science as a core subject and for the assessment of pupils' progress will make demands on the co-ordinator. Every school must account for the time spent on science and implement the programmes of study. This will require a review of existing practice, monitoring and evaluating the schemes of work resources and organisation. The assessment requirements make it necessary for the co-ordinator to develop moderation skills. Thus, in order to meet these demands, the co-ordinator will have to develop the evaluator function of their role.

Co-ordinator

The co-ordinator function of the role will need to be expanded because of the necessity to present an overall picture of science within the school not only for presentation to the LEA, governors, parents and other outside agents, but also to ensure that the programme of science for the children is a coherent one. Liaison with other co-ordinators in the school will be of increasing importance to ensure that appropriate cross-curricular links are made and avoid the fragmentation of the primary school curriculum. Furthermore liaison with secondary schools will be essential if the idea of continuity across phases is to become a reality.

The points made above are only some of the implications the introduction of the National Curriculum will have on the role of the science co-ordinator, there will be others. What is certain is that the demands made on the co-ordinator will not decrease, if anything they will increase, as Harrison & Theaker (1989) say:

> It is our view that there is no way a headteacher could hope to provide the detailed knowledge and understanding of each area of the National Curriculum even if s/he wanted to. . . . Thus the role of curriculum leader, post National Curriculum, will be more rather than less necessary, the scope is increased but, of course, so is the importance of doing it well from the outset. (page 6)

Obviously all the changes cannot happen at once because of the variety of factors which influence the performance of the co-ordinators' role, not least the personal qualities of each individual co-ordinator. However it seems that one crucial element is the training of the co-ordinators themselves who need to see their role in its wider context. They need to develop their own expertise not only in terms of their scientific understanding, but also in terms of their ability to carry out the managerial demands of the role. The proposed model provides a framework around which the training demands of co-ordinators can be identified and the effectiveness of the co-ordinator appraised.

Conclusions

The role of the science co-ordinator in primary schools is very complex. Each individual situation generates its own set of demands, constraints and pressures on the co-ordinator and every co-ordinator reacts according to his/her own personal experience and expertise. Thus any general conclusions that are drawn have to be seen in the light of such individual differences.

Although headteachers, co-ordinators and classteachers have a very similar perception of the role of the science co-ordinator, the commonly held view is somewhat restricted to the organisation of resources and the writing of a policy for science which are considered very important. Other aspects of the role, notably working alongside colleagues in their classrooms, are regarded as being considerably less important. This restricted view of the role gives rise to some concern because the co-ordinator will have to take on wider responsibilities if demands of the National Curriculum are to be met.

Meeting such demands and increasing the effectiveness of the science co-ordinator will depend on many factors, not least time, resources and the relationships between the co-ordinator and other staff. More fundamentally the way in which the role of the co-ordinator is perceived needs to be reassessed. The proposed model is presented as a step in this direction.

References

Anon (1987) An initial evaluation of LEA Guidelines. *Teaching Science* 5 (2), 5–9.

ASE (1981) *Science and Primary Education Paper No.3: A Post of Responsibility in Science*. Hatfield: ASE.

——(1988a) *Initiatives in Primary Science: An Evaluation, Report*. Hatfield. ASE.

——(1988b) *Initiatives in Primary Science: An Evaluation, The School in Focus*. Hatfield: ASE.

Bell, D. (1990) The role of science co-ordinators in primary schools in three local education authorities. Unpublished M.Ed. Thesis, University of Liverpool.

Blyth, W. A. L. (1965) *English Primary Education Volume One*. London: Routledge and Kegan Paul.

Campbell, R. J. (1985) *Developing the Primary School Curriculum*. London: Holt, Rhinehart and Winston.

Central Advisory Council for Education (England) (1967) *Children and their Primary Schools: The Plowden Report*. London: HMSO.

Clayden, E. (1989) The management of curriculum development in Primary Schools—theory and practice. *School Organisation* 9 (1), 27–38.

DES (1975) *A Language for Life: The Bullock Report*. London: HMSO.

——(1978) *Primary Education in England: A Survey by HMI*. London: HMSO.

——(1982a) *Mathematics Counts: The Cockroft Report*. London: HMSO.

——(1982b) *Education 5–9: An Illustrative Survey of 80 First Schools*. London: HMSO.

——(1983a) *Teaching Quality*. White Paper Command 8836. London: HMSO.

——(1983b) *9–13 Middle Schools: An Illustrative Survey by HMI*. London: HMSO.

——(1983c) *Science in Schools Age 11: Report No. 2*, Harlen, W., Black, P., Johnson, S., Palacio, D. (eds). London: HMSO.

——(1984a) Education support grants circular 6/84. In DES (1984) *Circulars and Administrative Memoranda*. London: HMSO.

——(1984b) *Science in Schools Age 11: Report No. 3*, Harlen, W., Black, P., Johnson, S., Palacio, D., Russell, T. (eds). London: HMSO.

——(1985a) *Better Schools*. White Paper Command 9469. London: HMSO.

——(1985b), *Science 5–16: A Statement of Policy*. London: HMSO.

——(1985c) *Education 8 to 12 in Combined and Middle Schools: An HMI Survey*. London: HMSO.

——(1988a) *Science at Age 11: A Review of APU Findings 1980–84*, Russell, T., Black, P., Harlen, W., Johnson, S., Palacio, D. (eds). London: HMSO.

——(1988b) *School Teachers' Pay and Conditions Document 1988*. London: HMSO.

——(1989) *Aspects of Primary Education: The Teaching and Learning of Science*. London: HMSO.

Great Britain—Laws and statutes (1988) *Education Reform Act 1988*. London: HMSO.

Harlen, W. (1985) *Teaching and Learning Primary Science*. London: Harper and Row.

Harrison, S. and Theaker, K. (1989) *Curriculum Leadership and Co-ordination in the Primary School: A Handbook for Teachers*. Whalley: Guild House Press.

Morrison, K. (1986) Primary school subjects specialists as agents of school-based curriculum change. *School Organisation* 6 (2), 175–83.

Primary Schools Research and Development Group (1983) *Curriculum Responsibility and the Use of Teacher Expertise in the Primary School*. University of Birmingham, Department of Curriculum Studies. In P. H. Taylor (1986) *Expertise and the Primary School Teacher*. Windsor: NFER Nelson.

Raper, G. and Stringer, J. (1987) *Encouraging Primary Science*. London: Cassell.

Rodger, I. A. (1983) *Teachers with Posts of Responsibility in Primary Schools*. Project funded by Programme 2 of the Schools Council, University of Durham, School of Education.
Stow, M. and Foxman, D. (1988) *Mathematics Co-ordination: A Study of Practice in Primary and Middle Schools*. Windsor: NFER Nelson.
Taylor, P. H. (1986) *Expertise and the Primary School Teacher*. Windsor: NFER. Nelson.
Taylor, P. (1989) A change in role: Curriculum co-ordinators in the Primary School. *Education Today* 39 (3), 50–3.

SCIENCE IN PRIMARY SCHOOLS: INNOVATION AND EFFECTIVENESS IN NORTHERN IRELAND

Frank Curragh and Patricia Diamond

School of Education, The Queen's University of Belfast, Belfast, Northern Ireland, BT7 1NN, UK

Abstract   This study uses an empirical qualitative survey of a cohort of primary schools in Northern Ireland to identify and classify the indicators of effectiveness which govern the implementation of a programme of science education. The data obtained is essentially curriculum rather than assessment orientated since the Common Curriculum in Northern Ireland, although similar in nature to the National Curriculum in England and Wales, has an implementation lag of one year. The indicators of effectiveness identified are aligned to a general matrix of performance indicators and show that the effective implementation of new programmes of study in science education are supported by six participating measures. The professional competency of teachers is of cardinal importance and is dynamically coupled with adequate early resourcing and the leadership potential of individuals and change agents in the educational system. In addition, the methodology used to teach science, the psychological set of schools determined by stated aims and policy outlines together with the level of pupil achievement are predictive of quality. Various components of these indicators are discussed as primary meaures of effectiveness.

Introduction

The selective system of education in Northern Ireland influences and in certain areas dominates the curriculum presented to pupils. Consequently verbal reason and arithmetic skills are emphasised from an early stage in the primary school sector (Sutherland *et al.*, 1986, 1989). In the past, many primary schools operated a timetable which allocated about half the available time to English and mathematics, which meant that the basic skills were emphasised. Residual time has frequently been allocated to other skills and subjects which in some cases are offered as electives. Prior to 1981 it was rather difficult to establish how much science was included in the other subjects category since very little research in science education was carried out in the primary sector of education. Nevertheless the Department of Education Inspectorate Report of 1981 estimated that, traditionally in the past, about 2.0% of primary schools offered some science as a rather narrow-based study of horticulture and rural science (Kerr, 1987). The 1981 report estimated that at this time about 10.0% of primary schools had a well defined programme of study in science which had an orientation towards the biological sciences rather than the physical sciences. There was little emphasis by teachers upon skills or processes, the programmes being characterised by content which was descriptive rather than conceptual in nature. Some of the reasons given in the report for this lack of esteem for science originated from poor resourcing, the weak scientific background of the teachers and the restricted value placed upon science in the curriculum of primary schools.

The decade between 1981 and 1991 reflected a substantial improvement concerning the esteem and structure of science education in primary schools in Northern Ireland. The report of the Assessment and Performance Unit (1984) (APU) revealed that 56.0% of primary schools had written programmes of study in the realm of science and that 40.0% of schools had appointed a science co-ordinator who had the responsibility of developing such programmes. Those teachers who acted as co-ordinators also assumed a degree of responsibility for training their colleagues in relevant aspects of science. This report also showed that pupils in the primary sector of education in Northern Ireland were among the highest achievers in the processes of science related to observing, measuring and recording. They were less effective in the realms of planning experiments, hypothesising, controlling variables and carrying out investigations, nor indeed was the range of curriculum topics taught in science as diverse as that found in England and Wales.

The period between 1986 and 1989 was one of fairly rapid growth in primary science. To a large extent this was due to the effective work of the Northern Ireland Council for Educational Development which published a series of guidelines for primary schools. One of these, Guidelines for Primary Schools Projects (Science) contained many aspects of science which were later exemplified in national programmes. For example, the guidelines in science suggested how programmes of study might be planned, outlined the nature of science and indicated how assessment, especially record keeping, might be carried out. A plethora of in-service courses resulted in a significant improvement in the quality of science taught in primary schools and an increased opportunity for teachers to gain qualifications in science. Up to this point the position of science in primary schools in Northern Ireland was very similar to that reported in the various surveys carried out in England and Wales (1978, 1984, 1989a and b) which suggested that science was slowly achieving a new status consequent upon the predictions of available reports.

Sweeney (1989) showed that primary schools in Northern Ireland were providing much improved opportunities for science and that the programmes of study used by teachers strongly resembled the concepts which were embodied in the national curriculum proposals. There was increased classroom practice in science with 60.0% of schools using written programmes of study and 25.0% of these specifying pupil skills which were carefully aligned to appropriate content areas. The regulatory policy orientation of the science guidelines, together with emerging evidence concerning the nature of the national science proposals were reflected in programmes of study which fairly accurately predicted what was later to emerge as the Common Curriculum (1990) in Northern Ireland. Sweeney also indicated a range of concepts, which included change of state and aspects of electricity, which teachers found difficult not only to understand personally but to teach and assess. Many teachers in this survey expressed a foreboding perigenesis of the assessment procedures embodied in current policy statements. While the preference of most teachers supported the practice of pencil and paper assessment procedures about 11.0% were engaged in some form of record keeping which was exemplified in the science guidelines. In many instances the programmes of study used or under development in schools were in sympathy with those illustrated within the science guidelines and mirrored fairly accurately the content of the emerging national curriculum schemes.

The publication of the Common Curriculum in Northern Ireland, which to a large extent is similar in nature and content to the National Curriculum of England and Wales, concominantly signalled a synergy of curriculum development within schools, Education and Library Boards (equivalent to LEAs in England and Wales) and educational institutions which also offered enhanced educational learning processes in science education. The present research was carried out against this background of activity, the dynamics of which, it was assumed, would produce effective improvements in the provision of science programmes of study in primary schools. Past experience indicated that the inherent nature of the teaching profession in Northern Ireland would primarily define the consequences of any educational policy in terms of pupil achievement rather than reflect any organised climate of ideological thought concerning the nature of a centralist policy which instrumentally designed the concepts and experiences of subject content. It was against the background of perceived rapid growth in science that this research was carried out and which Sweeney has shown to be sufficiently elegant to be worthy of further research.

Research Premises

The research was carried out between December 1990 and April 1991, dates which allowed about one year to elapse after the publication of the Common Curriculum Science document in September 1990. This might be considered a fairly short period of time in which to expect educational improvement especially since Northern Ireland lagged one year behind the publication and implementation of all the schemes pertaining to curriculum development in England and Wales. In addition, there appeared to be an unwritten policy in Northern Ireland which followed the views of Harlen (1983, 1989) which suggested that the development of programmes of study should precede any attempt to formally assess pupils. Consequently this research attempted to identify the performance indicators, within primary schools, which made facile the introduction of innovative programmes of study in science. It was envisaged that any conclusions reached would be of benefit to those developing and implementing new curricular ideas nationally and influence the structure, timing and nature of in-service courses provided by Education and Library Boards etc. In particular it was anticipated that those managing change within the curriculum of schools would benefit from some insight concerning the implementation process and the identification of some of the pre-requisites considered necessary to ensure quality within the educational system when new programmes of study were required by law to be hastily institutionalised.

A preliminary survey of some six primary schools indicated a persuasive case to maintain caution on the part of the researcher in this rapidly changing scene and in particular the area of science which was new in style and approach to most teachers. Evidentiary responses from teachers indicated a responsibility to show that the research was not involved with regulating or evaluating the educational system which was sensitive both then and now to academic elitism in science and hesitant to afford open access to classroom practice in science. Other insights from the preliminary survey revealed that teachers felt a degree of inadequacy, arising mainly from the lack of comparison with developments in other similar schools and that a syndrome of anxiety persisted within very diligent mature teachers.

Research Methodology

A qualitative research strategy was judged to be most profitable in the prevailing circumstances although it was anticipated that a systematic measurement of some requisites and pre-requisites could also be made to afford useful quantitative data (Gray *et al.*, 1990). Three aims regulated the study which employed a structured interview technique. This was used to differentiate as far as possible, hierarchically, the factors perceived by teachers as important in institutionalising new programmes of study in science. The aims encompassing a wide range of objectives were:

(a)   to identify the developments in science teaching in Northern Ireland which has resulted from the introduction of the common curriculum;

(b)   to construct a matrix of factors which would be regarded nationally as performance indicators in implementing and improving programmes of study in primary school science;

(c)   to suggest how those indicators distinctively influence the effectiveness of a school.

Interviews were carried out in a sample comprising some 130 teachers (the majority were principals of schools or science co-ordinators within schools), five representatives from Education and Library Boards and six science educators, some of whom had responsibility for designing professional award bearing courses in science. The teachers interviewed represented 75 schools (about 11.0%) in the primary sector of education. Schools were sampled by size and were proportionally distributed throughout each of the five Education and Library Boards. The significance of size of school was considered to be important since those with eight or fewer teachers had a principal who also taught thus reducing the administrative capacity within the school. The sample included controlled and voluntary maintained schools, integrated (by religion) schools and the whole sample was selected to represent the urban/rural dichotomy which exists in Northern Ireland.

Each interview lasted between 1.5 and 2.5 hours and was structured in method and content to reflect the positive achievements of the school in terms of implementing a programme of study in science. No direct attempt was made to focus upon gender issues nor was there an attempt made to locate the social milieu of the school. Field notes, observations and interview reports were analysed by three professional educators after each visit to characterise qualitatively the following issues and as far as possible to monitor these in quantitative terms:

(a)   the nature and coupure of science within the school,
(b)   the support systems inferring potential for curriculum development,
(c)   the means whereby written programmes of study are developed,
(d)   the advantages perceived by specific methods of teaching,
(e)   an analysis of teacher self-perceptions,
(f)   the allocation and use of resources,

(g) the professional development of teachers,
(h) the means whereby pupils are assessed in science,
(i) how scientific knowledge diffuses within the school,
(j) the teachers' view of science,
(k) the administrative requirements to implement science within the school organisational system,
(l) decision-making in school policy concerning science,
(m) the aims of establishing programmes of study in science.

It is accepted that this research reflects a rather tentative approach to identifying possible indicators for effective programme building in primary school science since some programmes were in initial stages of development and others enhanced productions of previous work. To conceptualise this whole field of actually measuring these indicators at least three issues emerge as importnat in Northern Ireland. As part of a global village of educational reform in the British Isles it has some unique conceptual structures such as, a recognisably professional and stable workforce and one which offers certain criteria of truth exemplified by sincerity and integrity within the workforce. Firstly, the selection procedure, which currently does not include assessment in science, could mistakenly be compared by teachers and parents, to the progressive differentiation by level envisaged by the Task Group of Assessment and Testing (1989). No significant unitary system of assessment has yet emerged in Northern Ireland and therefore any allusion to selection in this research was avoided and consequently summarily dismissed as a major measure of effectiveness in implementing programmes of study. It is anticipated that the formal assessment of pupils will follow by one year the policy enacted in England and Wales. Secondly, although schools in socially advantaged and disadvantaged areas were included in the sample, no specific attempt was made to identify any effectiveness indicators in science as parameters of socially deprived or favoured locations. Thirdly and closely related to the two above, no attempt was made to take into account the sectarian milieu which surrounds the school which could, if desired, define some aspects of effectiveness. It should be pointed out, however, that many schools in Northern Ireland have successfully structured their consciousness of operation to overcome alien external (to the school) conditions. Thus, the research is seen as the first phase of a three phase longitudinal study to monitor school effectiveness with respect to implementing programmes of study in science. Contextually the first phase of school effectiveness depends upon the inception and implementation of programmes of study. In this case the programmes of study are required as part of a national curriculum reform and represent the initial phase of a study sequence. The second phase, quantitative evaluation of these programmes requires more incisive classroom observation which in turn leads to the third phase. This latter phase involves various quantitative measures related to the level of pupil performance which will not be available before 1993.

Results and Discussion

It was interesting and indeed re-assuring to find that there was a stimulating work ethic in this sample of primary schools. Many teachers desired some external indicator of professional approval as a reinforcement of their status in science. Unfortunately this can only be meaningful if it is accorded from a position of external authority such as Education and Library Board advisers or members of Her Majesty's Inspectorate. There was no significant disparity found between the quality of work being carried out in controlled or voluntary maintained schools. It is significant that many integrated schools (a minority in Northern Ireland) reflected a less confident approach in science which might be attributed to inadequate early resourcing but now accorded similar status to other schools. This pervicacious proclivity is slowly dispersing as these schools have recently received government finance to support all of their activities. For the reasons outlined later in this article the size of school did not radically affect the development of science. However, the struggle to meet the demands of quality assurance was somewhat more apparent and acute in small primary schools where science was viewed as the most time consuming activity. An important factor, that of formal assessment of pupils, although operationally viewed as important by teachers, did not immediately impinge upon their consciousness. The reason governing this view is associated with the timing of the whole process of curriculum innovation in Northern Ireland which is about one year behind that in England and Wales. Thus the assessment of pupils was not a formal requirement at this stage nor are there precise techniques yet identified by the Northern Ireland Schools Examination and Assessment Council. These are likely to be different from those currently in use in England and Wales. Teachers consequently gave priority to the construction of a programme of study in science and left for another time the formal assessment of pupils.

The cohort of teachers in this sample reflected an objective rather instrumental approach to science teaching blending a descriptive explanatory paradigm with one strongly related to pupil process skills. It was clear that most of the modern challenges in science had been neglected in the plethora of in-service courses offered. Not only had the spirit of and introspection of Kuhn and Popper been excluded, but a niche, within future in-service courses, identifying the deeper concepts of science and a wider understanding of science was required. The unquestioning approach to science by teachers was also apparent when models of science curriculum were investigated. The central generation, control and focussed instrumentalism of the science content was seldom questioned within programmes of study. It might be argued that this generalisation is also found in secondary and tertiary programmes of study in Northern Ireland. Teachers did, however, illustrate a greater concern for quality assurance associated wth their work in science and especially how this would benefit their pupils. A characterisation of meliorism in the realm of primary science was acknowledged within the teacher sample rather than enhanced optimism with the progress they had made. It was also clear that the majority of teachers, who had already extensive knowlege of the process skills in mathematics, found it difficult to translate these to science. Nevertheless the results of the APU survey of 1983 indicated that the pupils were aware of these skills and could translate some of them to the area of science suggesting that teachers were effectively but unconsciously teaching such skills. Before considering the indicators of effectiveness which emerged when implementing the Common Curriculum science programme two issues merit discussion. The first relates to teacher perception of safety in carrying out scientific experiments. Many of these experiments were in fact quite safe and a highly trained professional would have recognised them as such and illustrated the need to seek an experiential approach to programme building in science. It is accepted that safety in science is related to manipulative competence and in many instances to the availability of appropriate apparatus or resources. Some teachers name a sink and electrical sockets as priority classroom requirements. The second important issue relates to teacher self-esteem. Many female teachers in the sample, with fifteen years experience or more failed to recognise that the programme building with which they were involved was, in fact, of a high quality. Lack of self-assurance was often portrayed in a rather frustrated, sometimes distressed, expression of dissatisfaction. This dichotomy does much to reduce the effectiveness of programme building and sometimes leads to a loss of confidence by other staff.

Some of the issues identified above mediate the performance indicators described below but are not necessarily related to any one of them.

Performance Indicators Aligned to the Implementation of Science in the Common Curriculum

The original notion of performance indicators in educational systems is strongly entrenched in economic and social models. In this study economic and personal professional characteristics emerge as important. Various attempts have been made to classify, in some conceptual form, the comprehensive range of indicators which are found within educational literature. For example, van Herpen (1989) lists a comprehensive overview of these, yet upon inspection they appear to be classified using six fundamental categories. These are:

(1) person-oriented indicators,
(2) resource-based indicators,
(3) social climate indicators,
(4) curriculum quality indicators,
(5) policy-related indicators,
(6) pupil performance indicators.

These categories may be sequentially correlated with those pseudo-hierarchically identified in this study as:

(1) professional development of those responsible for teaching science,
(2) resource availability for science,
(3) leadership potential,
(4) teaching strategies used to implement science programmes,
(5) school aims related to internal and external policy,
(6) pupil assessment.

These could be further refined and referenced in terms of input, output and process as described by Scheerens (1990).

The most significant indicator to emerge in the study, which focuses exclusively upon the implementation of science in the common curriculum, was that related to the teacher as a person in the educational arena. There is a temptation to ask 'what is teaching?' in the context of the range of administrative and other activities embedded in the Common Curriculum but this rightly belongs to a phase two evaluation of primary science programme building. Nevertheless, it seems that this value-laden, professional indicator is probably the most influential and persuasive indicator in determining the constructive implementation of programmes of study in primary science. The second most influential indicator, in what is essentially a fluid situation where one indicator diffuses only to be juxtapositioned to another, is resource based. That is, non-human resources related to equipment, books, space and other facilities where, within the context of expectations of primary schools in Northern Ireland, this is accepted as satisfactory. Third, in approximate order of merit, is leadership, both within and outside of the school. For the most part, the school principal acts as the authoritative change agent within the school although the science co-ordinator plays a very constructive, and in the majority of instances, a most effective role. External to the school, influential leadership was prosecuted by the Education and Library Boards and the Northern Ireland Council for Educational Development in arranging support teachers and creating a climate for innovative development. It is acknowledged that the term curriculum is related in the present context to educational activities and is an overarching one in terms of method, organisation and assessment. The fourth indicator of effectiveness falls within this category and may be identified through an analysis made of the methods used to teach science throughout the primary school and especially how the integrative curriculum structure operates within the school. In an ongoing and developing situation of science teaching the fifth indicator of effectiveness, 'school aims related to internal and external policies', is somewhat comprehensive and oversees the other indicators acting as a precursor of these. As an autonomous entity it has a profound influence on school organisation and management policy. The sixth indicator, related to pupil assessment and level of achievement, is currently embryonic in nature, awaiting definitive, reliable and valid strategies to be provided by the Northern Ireland Schools and Assessment Council. Nevertheless, schools are developing formal record keeping measures which profile pupil ability in science and are considered by many teachers to be the main means of assessing pupils in the primary sector. These are sufficiently well developed to afford optimistic teacher views concerning their ability to adequately monitor the performance of their pupils at whatever level of achievement. Many teachers feel, however, that to assess pupils formally at eight years of age is socially and academically devisive and would prefer that this policy be abandoned.

Each of these indicators is described in greater detail below. This research shows that there has been an enormous input of time, expertise and professional development to support the development of primary science. This reflects a system of science education in primary schools in a state of growth, improving in quality and demonstratively active in achieving the goals outlined by the Common Curriculum in Northern Ireland.

Professional Development

The teaching profession in Northern Ireland has been exceptionally stable over a substantial period of time. The low turnover rate in both primary and secondary sectors of education may be due to a tradition of teaching in families in Northern Ireland and the economic climate which does not offer extensive employment opportunities for new graduates. In this study the commitment and determination of the teaching staff to produce acceptable science programmes was conspicuously high in profile. The instinctive factor and major issue for many teachers was the future educational welfare of their pupils. Too frequently the self-perception of the teachers concerning the elegance of their programme design was tinged with unmerited insecurity. Most of the programmes were logically planned and valid, although not tempered at all times with a deep understanding of the scientific concepts involved. No significant gender difference was found in either the attitude to programme building in science or the ability of teachers to construct effective programmes. Nor was age a significant factor in determining the quality or self-esteem of teachers, a point somewhat at variance with the study of Carre & Carter (1990) who found that age and gender were influential facets of perception and esteem in science teaching.

In 19.0% of small schools and 25.0% of large schools there were teachers who either had a degree in science or some other formal qualification such as a validated unit(s) conferred through in-service programmes of study or certificates awarded by the Association of Science Education. It was interesting to find that 14.0% of staff were members of this organisation and that overall 42.0% of all teachers had acquired some form of qualification in science. These ranged from those listed above to attendance at summer school courses associated with science education. All of the teachers in Northern Ireland attend in-service courses arranged by the Education and Library Boards. In the sample of teachers interviewed, in addition to attending the general courses, all had attended similar specifically arranged science-based local courses.

Most (83.0%) of the teachers considered that science was a priority in the education of primary school pupils with 97.0% of all staff feeling confident that they could implement and teach science to a satisfactory standard. Those acting as science co-ordinators were in general accord with this view although the percentage was a little lower (91.0%) with some increased apprehension in smaller primary schools (88.0%). Nevertheless these high percentages reflect a profession significantly in control of the new structures. Their source of confidence was fairly equally distributed between the support afforded by the Education and Library boards, the leadership of the school principal, the collegiate responsibility of the staff and the esteem, self or otherwise, of the science co-ordinator.

Resource Availability for Science

In the past many science schemes in primary schools did not reach full maturity due to the lack of physical resources and in addition the poor quality of equipment provided was mentioned in the report of the Northern Ireland Inspectorate (1981). At that time 10.0% of primary schools were considered to have sufficient resources but 33.0% had no resources at all. Sweeney (1989) reported that almost all of the teachers complained about the lack of resources and considered this to be a major concern which adversely affected the implementation process. This study revealed that a dramatic change had taken place over the past two years which, for the most part, was due to a substantial financial input from the Education and Library Boards. This source of economic support was supplemented, in some instances by parent associations who purchased equipment for the school and in the case of integrated schools by charitable bodies. As a result effective resourcing of schools was claimed by 79.0% of large schools to be sufficient for whole class teaching. However, there was insufficient equipment to prevent sharing of resources by pupils and certainly inadequate stock to allow each class in larger schools to operate as an independent science unit. Non-equipment resources such as books appear to be adequate with 91.0% of all teachers satisfied with the range available within schools. Many teachers extolled the virtue of the loan service provided by the Education and Library Boards while 96.0% reported that there was adequate supply of reference books for pupils and teachers within schools.

There was a consensus of support in the Greater Belfast area for the very innovative idea of a science bus which was introduced and operated by the Ulster Museum. Prior discussion with the science co-ordinators in a school allowed a science adviser in the museum to develop specific science topics and bring these to the school where collaborative teaching ensued. A major resource, used by 90.0% of the teachers in the survey, reflected the effectiveness of current television broadcasts in science. This represented an increase of about 5.0% in the number of teachers using television broadcasts in 1989. In both this research and Sweeney's study of 1989 teachers commented positively upon the appropriateness of the broadcasts for both key stage 1 and key stage 2. In contrast, radio programmes in science were less popular and appeared to be of less value to schools in general. While science as an investigative classroom process was considered by many teachers to be difficult this was not the case when conscious use was made of the local environment of the school. For most schools this was the arena where teachers considered that true investigations were carried out by pupils working either singly or in groups. Virtually all of the schools made use of this resource together with the variety of field centres available to them. In addition, excellent programmes of study in science were available in the planetarium, museums, science fairs and aquaria. The difference between the plethora of activities undertaken by most schools in this survey is in direct contrast to the 1981 report which indicated that apart from a nature table the resource nature of the playground, parks, seashores and forests were largely ignored by schools.

All of the teachers in this sample drew attention to the future development of science in their schools. The relatively new, and for the most part, untried local financial system of resourcing each school was cited as a source of concern. While it is impossible for teachers to identify the precise policy indicators, in either a general or particular way which might adversely affect science, it was assumed that capital development schemes might be jeopardised and technical support an unrealistic dream. Other concerns also related to financial control identified the need to reduce class size in order to carry out investigative work in science, the possible reduction in travel to field centres and the purchase of additional equipment and the replacement of old stock. Of these, class size and safety dominated the persuasive harmony of most interviews. Teachers in general expressed the view that, for science, class size should be around 24, to accord with the safety regulation operating in the secondary sector where additional technical support is fairly readily available.

Leadership Potential in the Educational System

The third indicator of performance is again person-oriented and, although financial in orientation facilitates the person as a professional. This originates from the policy operated by the Education and Library Boards in Northern Ireland. It is quite difficult to accord any degree of priority or indeed hierarchical order to the leadership given by individuals such as the principal of a school, the science co-ordinator or support teacher nor is the collective responsibility of colleagues, in-service adviser groups or support teacher commitment easy to quantify. All of these entities fuse to provide leadership and support which is essential to the internal harmony of a school embarking upon designing and implementing new programmes of study in science. Through an adviser system the Education and Library Boards provide differentiated support to assist schools implement the science curriculum which is difficult to sequence and is unnecessarily diffuse with respect to the range of targets posited within it. Each of the Education and Library Boards have recognised the difficulties confronting small schools and in a concerned way provided support teachers to help establish programmes of study in science. Within this sample 61.0% of small schools received such support on a yearly basis and 19.0% of the larger schools were accorded support for one term. This type of support is seen by many teachers to be very useful and significantly better than attending courses since the former is primarily intended to cater for the specific needs of the school in question whereas most courses have a tendency towards the general. In many instances support staff have helped develop programmes of study but the most effective commitment of this group concerned the reassurance and confidence building processes they effected with teachers. The assumption that it was only the small schools which required this type of support was found to be erroneous in some specific cases. Many of the schools with up to ten teachers on the staff and where an effective science co-ordinator did not exist, often reflected a distinct lack of staff confidence. This group would certainly benefit from the presence of a support teacher. For similar reasons many of the larger schools would benefit from more prolonged support than board policy allowed.

In 79.0% of the schools the presence of a science co-ordinator is seen as significant. In general they promote course development and stimulate other teachers. Thus in the primary sector of education there is an emerging developmental structure which has grown substantially since 1988 when only 43.0% of schools had developed a structural dimension in science. Many (94.0% in large schools and 54.0% in small schools) of the science co-ordinators had held the position for over one year and in some larger schools (29.0%) there was a separate co-ordinator for key stage 1 and key stage 2 which added cohesion and solidarity to the school system. In such cases there was close collaboration between the two people involved but the enduring view was that a single co-ordinator for science throughout the school was pedagogically more effective in both small and larger schools. The science co-ordinators were the first group targeted for in-service training by the Education and Library Boards. They, in turn, provided some school-based in-service training for other members of staff. This required them to possess not only an acceptable personality and powers of communication but to possess the necessary scientific knowledge to carry out the tasks demanded of them. In many instances the principal played a very important 'change agent' role and in small schools acted as sole 'change agent' not only for science but all other subjects. Both the principal and the science co-ordinator made excellent use of days of exceptional closure and of directed time to accomplish the 'routines' of programme development applicable to each age grouping within the school. In some cases the principal of the school organised the school timetable in such a way that the science co-ordinator either taught or helped team-teach science classes throughout the school.

The leadership given by the field officers and advisers from the Education and Library Board was seen as constructive by teachers who in some cases expected specific programmes for direct school implementation to be provided by this group. While this approach was impractical and limited in use some of the resource materials, produced in booklet form by advisers, was elegant in construction and most useful especially when teachers began to align content to levels of pupil achievement. This type of provision was not uniform throughout each of the

Education and Library Boards and it is anticipated that this and other principles of co-ordination between the 'boards' will be encouraged by the recently established Regional Training Unit. There is little doubt that the systematised approach to in-service training has resulted in establishing very effective programmes of study in science in the primary sector of education. In particular the in-service courses operated by various educational institutions were very well received by teachers while all welcomed the provision of a new degree course specifically aimed at developing science for those hoping to enter the teaching profession.

Teaching Strategies Used to Implement Programmes of Study in Science

The fourth indicator of effectiveness in deciding if schools have satisfactorily implemented harmonious curriculum structures in science is process based. The elements of significance in this indicator relate to confidence in style of teaching used by the teacher. Those using integrated approaches implied that excellent programme development had resulted in a subordinate replete methodology consistent with excellence. For the most part the processes reside within the organisational structure of the school and therefore depend upon the decisions of senior management teams. It was evident that increasing numbers of teachers felt confident to integrate the pedagogy required in science with that demanded in English and mathematics as compulsory core subjects. As a result a highly satisfactory milieu of programme implementation and teaching had been achieved in a significant number of schools. In the present sample 50.0% of teachers operated in this cross-curricular way with 47.0% in small schools. The majority of teachers (67.0%), however, preferred a strategy whereby science topics only were taught, integrating these where relevant and somewhat partially with the compulsory cross-curricular themes such as health education, information technology and aspects of language. At appropriate points, use was made of graphs, classification diagrams etc. to combine rather than integrate the study of science with other areas of the curriculum. This group of teachers reflected a consistent but moderate level of achievement. In contrast to this fairly confident group of teachers there was a more hesitant group who taught science in discrete units within a specific lesson which they referred to as the science lesson. Overall there were 19.0% of teachers in this group with 21.0% in small schools. All school programmes of study included some cross-curricular identity within the aims of the study but in presentation the degree of commitment to these was somewhat limited.

It was pleasing to note that 37.0% of the total teacher sample attempted to employ a group-based semi-investigative approach to science studies although in the smaller schools, which had to contend with vertical pupil groups, teachers found this approach to be difficult although a corollary of their system. About 26.0% of teachers employed in small schools used this teaching strategy which although influential in style must be considered numerically small. In sharp contrast 68.0% of teachers in small schools used a combination of methodological approaches to science teaching which was slightly higher than those who favoured this approach in large schools (53.0%). It was pleasing to note that many teachers (37.0%) favoured group work in science as opposed to teaching the whole class which had few advocates (3.0%). There is evidence to suggest that group work in small schools is fairly difficult to organise and is especially so in very small schools.

Teachers appear to favour and pursue where possible a structured approach to science teaching with recognisable attempts to employ different strategies. There is, nevertheless, a fairly low level of effectiveness in this domain and teachers frequently argue that this is due to the constraints of space and the restricted availability of sufficient sets of equipment.

Teacher use of and attitude to allocated time for science also gives some indication of effectiveness although only indirectly related to achievement. While most (76.0%) teachers spend about 2.5 hours per week on science related teaching some (21.0%) feel that this is in fact over allocation and that some of this time should be given to other compulsory core subjects. About 10.0% of teachers in all schools indicate that the amount of content contained within the set targets requires a greater time allocation. Perhaps it is the views of the majority of teachers which are most appropriate in this context. The vast majority agreed that science should be part of the curriculum of primary schools and that a 10.0% allocation of time to study this area was about correct, however distributed. In this sample of schools both the allocation of time and the methods used to teach science are measures of school effectiveness in both large and small schools.

School Aims Related to Internal and External Policy Statements

The fifth indicator is input-based and illustrates the state of readiness of schools to innovate and implement programmes of study in science. A statement of aims, and the creation of this statement to a large extent represents a clear view of school policy. Therefore the schools which had extensive, often criteria referenced, aims and objectives were most effective in implementing programmes of study. While all schools had a set of general aims for the school, 84.0% had some form of written policy statement for science. An important catalyst for this component of effectiveness originated with the publication in 1986 of the Primary Guidelines Project (Science) by the Northern Ireland Council for Educational Development. Apart from creating a climate of readiness this document suggested a variety of principles fundamental in nature but extremely useful when exploring the design of a new curricula. To a large extent the guidelines gave a degree of psychological set to primary schools and encouraged unhindered programme development in schools. Many schools took this early opportunity to develop programmes of study in science and some included pupil skills. Probably of greater importance was the co-operative spirit developed within schools and the ownership of knowledge which resulted from group interaction. Indeed, many teachers prefer this means of characterising science rather than operationalising an imposed curriculum.

The rate at which programmes of study were developed is seen as an important indicator of effectiveness. Previously known data from 1984 suggested that 56.0% of schools have some form of documentation, while in 1989 the figure rose to 60.0% and is currently around 84.0% with 76.0% of programmes containing the process skills to be undertaken with pupils. It may be assumed that many of the ideas currently being developed first emerged within schools in 1987. Without these background documents it is doubtful if progress would have been so consistent or sophisticated. All teachers had heard of the guidelines and 61.0% had actually used them to generate specific aspects of the learner environment.

Although the aggregated number of schools having well documented, cross-referenced programmes of study in science is satisfactory, the small schools lag somehat behind the larger ones with 79.0% of all schools either having or in the process of developing such programmes. Unfortunately about 16.0% of all schools surveyed had no written policy statement dealing with science. Apart from suggesting a degree of slothfulness and possible edentate leadership within the school or pre-occupation with other subject areas there are no economic or resource reasons to suggest why this should be so.

Those schools which had some tradition of teaching science found it easier, than those not having any experience, to facilely accommodate the requirements of the Common Curriculum. Many of the teachers in the former category suggested that the target areas were over prescriptive and over elaborate with respect to explicit content and the compulsory cross-curricular themes defined within the Common Curriculum.

Pupil Assessment

This indicator of effectiveness is a summative measure of effective programme planning and authoritative teaching. Pupil level of achievement is seen by many researchers in the field of effectiveness as the most important of the output indicators, yet in raw terms, it says little of the socio-economic conditions influencing both the school and the pupil. There is little evidence to suggest that any of the indicators of effectiveness described above will be employed to modify the raw scores associated with the level of pupil achievement. Indeed in Northern Ireland some influential indicators of effectiveness might be difficult or impossible to quantify. For example, school location and the effects of long term unemployment within families are relevant entities on any scale of measurement. In this survey, apart from the outright condemnation by teachers of publicly declaring results by school, there was a feeling that school status depended upon the number of pupils qualifying for grammar school education. It is quite possible, that however the levels of achievement are expressed, this perception of effective schooling will remain. Teachers felt that the proposals made by the Task Group on Assessment and Testing (TGAT) were the most radical of all the reforms encountered. There was expressed relief that the formal testing of pupils would follow (by one year) the systems currently operating in England and Wales and that the procedures used would be somewhat different from the partially discredited standard attainment tasks. It seems that testing will arise from more normal classroom interactions in Northern Ireland and that supplementary resources will be available to help teachers assess pupils. These will not deliberately contrive practical situations within which testing will take place. Teachers have aleady recognised that there is evidence of enhanced pupil experience, motivation and self-confidence within the investigative context.

Sweeney (1989) reported that any form of organised assessment of pupils in science, especially process skills, is rare in the primary sector of education in Northern Ireland. This conclusion is disappointing since the available evidence nationally suggested that about equal weighting would be given to knowledge and process skills in emerging assessment schedules. The overall percentage of teachers in this sample who actively engaged in assessing pupils was 20.0%, of which about 9.0% used direct observation of pupil work, 6.0% used written tests and 6.0% used pupil work schedules. Although teachers illustrated a degree of confidence in their ability to conduct standard assessment tasks and to mark or grade such assessments they felt a distinct lack of confidence when asked if they could devise relevant assessment tasks. To test teacher ability, Sweeney devised three process orientated tasks for classes in the fourth year of primary school. Teachers found that it was very difficult to separate the assessment of content, which was frequently norm referenced, from the assessment of pupil skills which are criterion referenced.

In the present study teachers felt that they were not assessing pupils effectively or correctly since very few were able to place them in any level of achievement graded on a 1–10 scale suggested by TGAT. However, when prompted it was apparent that many teachers were currently engaged in developing some forms of assessment mostly related to profile schedules. Overall, 54.0% of teachers were in the process of developing some techniques of assessment with no significant difference in the level of activity in small or large schools. The majority (47.0%) were involved in developing record keeping on card or computer anticipating that this format might be the preferred means of assessment identified by the Northern Ireland Schools and Assessment Council. Although 61.0% of teachers expressed a degree of apprehension concerning the administrative arrangements for assessing pupils, they had a fairly clear idea of the techniques they would prefer to employ with some advocating and supporting a mixed frame of reference. The majority (61.0%) thought that informal teacher perceptions of pupil ability would be best, especially if teachers were involved with teaching the same group of pupils for periods of three years. Oral questioning was esteemed by 47.0% of teachers and 30.0% felt that practical work would be the best technique. Written assessment scheules were surprisingly infrequent with only 31.0% supporting this approach to assessment while 23.0% favoured the assessment of pupil project work as a means of reporting achievement.

As an indicator of effectiveness, assessment of pupil performance will remain a very controversial one if a selective system of education continues to operate in Northern Ireland. The results of this study suggest that an extensive range of school-based development has yet to take place in the area of pupil assessment. Teachers are not oblivious to the problems concerned and are already focusing upon and reinforcing their approach to pupil assessment. It is apparent that this indicator of effectiveness will continue to be a significant and controversial one and in the absence of definitive techniques of assessment being available to teachers it will increase the frustration threshold within schools.

Conclusion

The results of this study are reassuring to the extent that they reflect a system in growth and improvement, however mandated. Component data associated with the indicators of effectiveness explicit within the research show that the professional competence of teachers of science is moderately high and growing. Structure within the field of science education in primary education is changing through the recognition of work associated with the roles of the science co-ordinator and the support teacher to which one must add, in terms of effectiveness, the importance of prior planning and resourcing. The formal commitment of resources, adequate for whole class teaching, by the Education and Library Boards have constructively enhanced the economic homogeneity of the primary school sector and added to the sense of security and dimension of success felt by many teachers. Although the range of teaching strategies used by teachers of science in primary schools are somewhat limited, this, teachers argue, stems from the large number of pupils per class and the lack of multiple sets of apparatus. Few teachers had a deep understanding of the scientific concepts taught nor was there extensive knowledge of norm-referenced or criteria-referenced methods of assessment. While it was encouraging to note that many teachers were making worthy attempts to improve the techniques used to assess pupil levels of achievement many were adapting a policy of procrastination preferring to review the methods, as yet unknown, being developed by the Northern Ireland Schools Examination and Assessment Council.

The indicators of effectiveness described above conspicuously emerged as the study progressed and can be satisfactorily measured using component data. It was unfortunate that pupil achievement data in science is not yet available and in any event merits a specific study at a later date. Although many primary schools are referenced by a poor socio-economic climate and others additionally placed close to areas of sectarian strife, they manage to retain a remarkable academic homogeneity and sense of purpose. Nevertheless raw output scores of pupil achievement must be seen in the context existing around and within such schools and in such circumstances, this study indicates, there are many problems associated with reliability and validity of component measures. To quantify such factors requires additional research which would include measures of truancy rates and degree of managerial order within schools. It is clear that many schools have already advanced to reach new levels of achievement which may well embrace other contextual indicators not described in this study.

The indicators identified in this research arise within an evolving system of curriculum change and while logical in construction, measurable in terms of school performance and mutually competitive, more sophisticated research is required to equate them to the teaching and learning situation. It is rather difficult to describe them in terms of efficiency or 'value for money' since it would require detailed interventionalist policies on the part of education and library boards to compare financial arrangements between schools and in correlating past and present spending within a school. Gray (1990), discussing some of the issues, draws attention to the implicit aims of educational institutions and compares these to the single profit aim of commercial institutions and suggests that schools should not necessarily accept the priorities others would impose upon them. Tymms (1990) also recognises the imprecise science of inferring quality to systems in evolution. The unpredictability of inteventionist strategies is stressed and parallels drawn with indicator systems within school environments and in nature. In contrast to the indicators identified in this research Tymms discusses an alternative approach which uses the A-level Information System (ALIS) to determine the effectiveness of a school which means that the many facets of school effectiveness are to date only partially connotated.

References

Carre, C. and Carter, D. (1990) Primary teachers' self-perceptions concerning the implementation of the National Curriculum for Science in the UK. *International Journal of Science Education* 4, 327–41.

Department of Education Northern Ireland (1981) *Primary Eduation.* Report of an Inspectorate survey in Northern Ireland. Belfast: HMSO.

Department of Education and Science (1978) *Primary Education in England.* A survey by HM Inspectors of Schools. London: HMSO.

——(1984) *Science in Schools Age 11*. Report No. 2. London: DES.

——(1989a) *Science in the National Curriculum*. London: HMSO.

——(1989b) *National Assessment. The APU Science Approach*. London: HMSO.

Gray, J. and Jesson, D. (1990) The negotiation and construction of peformance indicators: Some principles, proposals and problems. *Education and Research in Education 2* (2), 1–16.

Harlen, W. (1983) Basic concepts and the primary/secondary science interface. *European Journal of Science Education 5* (1), 25–34.

——(1989) The national curriculum in science: Key stages 1 and 2. *Primary Science Review*. Special National Curriculum Supplement. Summer, 1989, pp. 10–12.

Herpen, M. van (1989) Conceptual models in use for educational indicators. Paper for the Conference on Educational Indicators, San Francisco.

Kerr, J. F. (1987) The development of the teaching of science in Ireland since 1800. PhD Thesis, Belfast: The Queens University.

Scheerens, J. (1990) School effectiveness research and the development of process indicators of school functioning. *School Effectiveness and School Improvement 1* (1), January 1990.

Sutherland, A. E. (1989) Selection in Northern Ireland: From 1947 Act to 1989 Order. *Research Papers in Education 5* (1), 29–47.

Sutherland, A. E. and Gallagher, A. H. (1986) *Transfer and the Upper Primary School*, Report No. 1. Northern Ireland Council for Educational Research.

Sweeney, J. J. (1989) Assessment in primary science. A study of problems in its implementation at age 8. MEd Thesis, Belfast: The Queen's University.

Tymms, P. B. (1990) Can indicator systems improve the effectiveness of science and mathematics education? The case of the UK. *Evaluation and Research in Education 4* (2), 61–73.

DEVELOPMENTS IN PRIMARY SCIENCE: A NEW ZEALAND PERSPECTIVE

Fred Biddulph and Malcolm Carr

Centre for Science and Mathematics Education Research, University of Waikato, Te Whare Wananga o Waikato, Private Bag 3105, Hamilton, New Zealand

Abstract The results of a New Zealand research project which sought to overcome major difficulties in primary science education are summarised. These include the development of a teaching approach which gives children's questions priority in investigations, the formulation of a set of evaluative procedures that are congruent with the underlying constructivist perspective, and the development of procedures aimed at changing teachers' beliefs and practices. Issues relating to the purpose of primary science education, children's learning, and teacher education are discussed.

Introduction

In the early 1980s primary science education in New Zealand, as in various other countries[1], had some major problems—despite recent curriculum development and other attempts to improve the system. These problems, summarised in Biddulph, Osborne & Freyberg (1983) include:

(i)   many teachers lacked confidence in the subject because they viewed science as a body of external truths which were difficult to understand,

(ii)   little science was being taught in primary classrooms,

(iii) children who did receive some science instruction were frequently bored by it, and

(iv) children often constructed unintended ideas from the transmitted science information.

Through no fault of their own, primary teachers were clearly in a bind with respect to science education. Their prior experiences (ironically in a positivist paradigm) had given them a negative view of science. Furthermore the accepted approach to teaching in science was behaviourist. The result was a disservice to primary science education and a disservice to science. The Learning in Science (Primary) Project (1982–84), funded by the New Zealand Department of Education and based at the University of Waikato explored ways of overcoming these difficulties. Subsequent major research projects have developed and enriched that investigation into the teaching and learning of primary science.

"

A perspective on primary science education

Since there was little consensus in New Zealand among teachers and science educators about the purpose of primary science education, the Project members sought a goal for primary science education. Such a goal should do justice to science, to children's learning, and also make sense to teachers. Osborne, one of the Project directors, was instrumental in constructing a powerful, but elegant, perspective, summarised in Biddulph, Osborne & Freyberg (1983). Essentially it states that the main point of primary science education is to help children make better sense of their world.

Translating a perspective into practice

It is one thing to propose an acceptable goal for primary science education; it is quite another for teachers to translate this into practice. The gulf between the idea and appropriate teaching strategies can be too great. The next phase of the research therefore focussed on devising a teaching approach consistent with current views of science and with recent insights into how children learn. The approach needed to reflect the reality of the classroom, and to help teachers develop confidence as science educators. Symington's (1980) Australian research into children's ability to see scientific problems in everyday phenomena suggested a possible way forward. Considerable research was undertaken into the ability of New Zealand primary school children (mainly aged 7 to 11 years) to ask investigable questions. From this came an alternative teaching approach and explorations of teachers' ability to adopt the approach, and of its effects on both children and teachers. These matters are the focus of the rest of this paper, together with a consideration of the appropriate assessment of the teaching and learning approach.

Children's Questions

The researchers were aware of the difficulty of developing a viable primary teaching approach based on children's questions since previous attempts such as the Nuffield Junior Science Project (Crossland, 1972; Lucas, 1974; Wastnedge, 1967); Science 5/13 (Henry, 1976; Science 5/13, 1972); the Elementary Science Study (Shymansky, Kyle & Alport, 1982) had not met with conspicuous success. Furthermore, as Biddulph (1989) noted in New Zealand, many primary school teaching practices actually inhibit question-asking by children, and Dillon (1988) despaired that United States teachers would ever be able to promote pupil questioning and inquiry because the habit of the teacher asking questions of pupils was so ingrained. Nevertheless a procedure to encourage children's questions was developed since there are sound curriculum, pedagogical, philosophical and psychological reasons why these should be central to their investigations in science (Biddulph, 1989; Biddulph, 1991; Biddulph, Symington & Osborne 1986). These include:

(i)   providing teachers with a guide to content that is most likely to engage children,

(ii)   offering insights into children's thinking,

(iii) encouraging children to use a process which is at the heart of scientific endeavour, and

(iv)   enabling children to perceive real purpose in their learning, to make better sense of their world, and to develop autonomy as learners.

The New Zealand Learning in Science (Primary) Project found that, after children engaged in some unguided explorations of a variety of topics, it was relatively easy for both researchers and teachers to elicit questions from them. This included some topics that children said initially did not interest them (Biddulph, 1989; Biddulph, 1991). 854 questions were elicited from children in 48 primary classrooms, giving a mean number per class of almost 18. Of these, about two-thirds were judged to be investigable by children, given teacher guidance, either through hands-on activities, or by consulting resource materials and knowledge-able people. Further, classroom teachers who witnessed their children asking questions were surprised at the hidden intellectual ability of some, and were positive about the prospect of using the children's questions as a basis for science investigations. Other findings from this phase of the investigation were:

(i)   For any topic, a number of common questions were asked by the children. Some New Zealand 10-year-olds on the topic 'rocks' asked 'What gives the rock its colour?' while their counterparts in Melbourne, Australia asked 'How did they [pieces of rock] get their colour?' The implication is that, with appropriate research, teachers can be forewarned of likely questions about a given topic.

(ii)   The children accepted as worthwhile questions asked by other children, and even adults, provided that they were in the children's own language. This suggested that a teacher in the guise of a senior investigator might be able to include in a study some important questions that the children had overlooked.

(iii) Many of the children's questions focused on aspects of a topic highlighted by the exploratory activities. It appeared from this that, by judicious selection of such activities, teachers could guide children in a subtle, but seemingly natural, way to consider key aspects of a topic.

(v)   The children were often able to suggest a range of possible answers to questions asked by their peers. When the question was 'Why do we need a skeleton?' some 9-year-olds responded that 'It is like a frame to support your body from just being a lump of skin on the floor' and 'Well, it's kind of like a roll cage in a stock car; it protects you'.

These various findings suggested that a teaching approach making use of children's questions had considerable potential and would be worth developing.

Interactive Teaching Approach

The first outline of an alternative teaching approach that accorded children's questions a significant role was sketched by Osborne in 1983. This was sub-sequently tried out by a beginning teacher with some expertise in science education in her own classroom, and by one of the Project members in four other classrooms. Some changes were made to overcome minor difficulties encountered, but the major insights gained were:
(i)   the approach suited children—they participated intelligently and enthusiasti-cally,
(ii)   it was manageable from a teaching point of view,
(iii) most teachers would be likely to need considerable advice about how to implement it with specific topics.

The approach evolved into what has become known as the Interactive Teaching

Approach (see the monograph by Biddulph & Osborne, 1984, a summary in Biddulph, 1991, and Biddulph, 1992). Briefly, this consists of five components and provides six roles for the teacher. The *components* are:

- a preparation time for the teacher,
- exploratory activities to raise questions in the children's minds,
- eliciting children's questions (and prior ideas),
- assisting children to devise and carry out investigations,
- helping children reflect on and share their findings.

The *teaching roles* are as

- a stimulator of children's curiosity,
- a challenger of children's ideas,
- a resource person,
- a senior co-investigator,
- a provider of a non-threatening classroom climate,
- a co-evaluator.

The researchers recognised early that, since the teaching approach was so different from what parents and other members of the community expect, it was necessary to include in the monograph for teachers a model letter for parents and others explaining the new approach. This letter pointed out that traditional teaching approaches had serious shortcomings, and that the alternative approach would make extensive use of the children's own questions, and investigations based on them. Children would be challenged to consider the views of others alongside their own, and investigative strategies developed by other students. The letter also indicated to parents that they should not be surprised to find their children beginning to enquire in relatively complex ways about the world around them. In this context it is worth noting subsequent comments that children were indeed talking about science at home and astonishing their parents with the range of their developing ideas.

Teacher Change

The trials outlined above, the action-research of the Learning in Science (Primary) Project, and its outcome the interactive teaching approach, had been conducted by a teacher and researchers whose perspectives on primary science education reflected a constructivist view of learning. Westbury's (1983) question of whether it was possible to convey an innovative approach to teachers by cost-effective, written materials alone was next felt to be a key one by the researchers. Trials of a unit on 'Floating and Sinking', which incorporated the interactive teaching approach, were therefore undertaken with two groups of teachers. The first group had no particular interest in primary science (and would be typical of the majority of primary teachers), the second group had a relatively strong interest in science education and were child-centred in their teaching philosophy.

Teacher trials

The results of the trials with the first group of teachers are summarised in Appleton, Hawe, Biddulph & Osborne (1984). Briefly, the teachers were unable to construct the intended meaning from the written materials, even though they thought they had, and were therefore unable to implement the approach as envisaged by the authors. Their approach was influenced by their prior ideas of science, of their role as teachers, and of how children learn. These factors were illuminated by classroom observers who both witnessed the teaching of the unit in the classrooms and interviewed the children and teachers from time to time.

In contrast to the above trials, those carried out by the second group of teachers fulfilled the researchers' expectations (Biddulph, 1989). It seemed that a minority of teachers could interpret the written materials and implement the interactive teaching approach as intended. Clearly, most teachers would require a special inservice course to support them in a change of perspective and practice.

Inservice course

Members of the Learning in Science (Primary) Project devised and conducted a week-long inservice course in 1984 to introduce interactive teaching to a group of primary teachers who had been nominated by science advisers as being potential resource people in primary science for their schools. Mindful that most of the teachers' prior formal learning had probably not been learner-centred, Osborne (the course director) and Biddulph endeavoured to develop the course on constructivist principles and to provide course members with learning experiences and an atmosphere consistent with this. A description of the course and details of the results of the teachers' subsequent work with children are contained in Biddulph (1989) and summarised in Biddulph (1991). In brief, the course identified participants' prior understandings of primary science education, enabled them through active interview and involvement to explore their own and children's views (neither always being scientific) about a topic, provided them with first-hand examples of an alternative interactive teaching approach in the classroom, and challenged them to adopt the components and roles as they felt able. A follow-up study suggested that almost all of the teachers were able to adopt the interactive teaching approach as intended. Teachers encountered some difficulties adjusting to the new teaching roles and acquiring sufficient relevant resources. They also felt external constraints and perceived pressures to conform to traditional practices (especially in evaluation); nevertheless both teachers and the students felt very positive about the approach. The teachers became aware of a greatly increased involvement and development of their children in science. The teachers also became more confident about taking science because they no longer felt they had to be experts themselves.

The issue of lack of suitable resources is beginning to be addressed by a continuing action-research programme involving a group of primary teachers in the Waikato region who have expertise in interactive teaching. Units for teachers are being developed (Biddulph, 1992) and books for children, which address their questions and are designed to challenge and extend their ideas, are being written.

Teacher development courses are also currently the focus of a major research contract from the New Zealand Ministry of Education. Called the Learning in Science Project (Teacher Development) this research is outlined in Bell, Kirkwood & Pearson (1990) and addresses the adoption, by both primary and secondary science teachers, of a more learner-centred teaching approach. An interim report on the findings and on planned research is found in Bell & Pearson (1991).

Preservice primary science education

The principles underlying the structure of the inservice course were incorporated into some preservice courses for primary teachers at Hamilton Teachers College[2], and the effects on preservice teachers investigated. The results are reported in Biddulph (1987, 1990). Although this was not a comprehensive investigation, the indications were that the students, almost without exception, were positive about the opportunity to experience an entirely different way of learning, to begin to overcome their anxieties about science, and to work with children in a much more interactive way. It was also clear that most students needed to participate in at least two supportive science education courses (each of 36 hours contact time) to enable them to change their views about science and children's learning, and to develop sufficient skills to work reasonably confidently with children. This is perhaps not surprising as the shift in perspective, and the sophistication of strategies, required by most people to use an interactive teaching approach effectively is considerable.

Evaluation

The interactive approach poses another challenge for teachers, the evaluation of learning in this new procedure. Rather than determining whether children have reached pre-specified objectives the focus, in keeping with a constructivist stance, is on identifying and evaluating the actual sense that children make of their experiences in science. This should include affective outcomes. As the teacher role of co-evaluator implies, the children themselves engage in self-evaluation as part of the interactive approach. The good reasons advanced for this learner-centred evaluation are to enhance metacognition and to help learners develop autonomy.

The education system in New Zealand, as in other countries, is undergoing many changes including the development of curriculum achievement targets, and procedures whereby teachers are required to measure progress through these. A serious problem with this approach is that evaluation and assessment procedures may too narrowly concentrate on knowledge outcomes. Since assessment crucially affects teaching, this will result in teaching towards rote-learning of facts. In New Zealand a draft syllabus for upper primary and lower secondary schools had been developed which was innovative and forward-looking in its view of learner-centred activities in the classroom, and of contexts for learning which ranged over wide interests for the learners. The present move towards a more structured assessment of learning has resulted in this draft document being side-lined, and the direction it promised (which had been widely welcomed by teachers) losing focus.

A recent Task Group on Science and Technology Education in New Zealand surveyed the needs of the workplace and of the community and from this a clear picture of the expected results of teaching and learning emerged. This was that a broader range of skills than is the current target was expected, including both oral and written communication, group-working capabilities, problem-solving capacities and learning to learn. Unless evaluation procedures are developed for measurement of these skills they will be neglected in teaching and learning while narrower skills such as recall of knowledge are emphasised in the assessment process. There is clearly an urgent need to develop and research assessment in these areas. The interactive teaching approach points towards some broader evaluation possibilities which will now be discussed.

Evaluative criteria

A set of criteria which suggest worthwhile learning in science is set out in detail in Biddulph (1992). They relate to the development of science understandings, to intellectual development (including reasoning ability), to the development of learning, investigatory, technical and communication skills, and to the development of personal and social qualities. For example, in our experience, successful learning is occurring if children demonstrate development (i.e. extension, reorganisation, or change) of ideas during a study, movement towards important current scientific ideas about a topic, an ability to give sound reasons to support their views, a willingness to ask substantial questions and to take responsibility for devising and carrying out investigations, and a propensity to make sense of their new ideas by relating them to prior ideas and experiences. Since the teaching approach encourages communication between students and between the student and the teacher about the developing learning there is ample possibility for monitoring communication skills and group interaction skills. Clearly there is a match between the desired skills described above and the outcomes of the interactive teaching approach. It would be regrettable if any reorganisation of science teaching and learning in the primary schools ignored the very positive contribution possible from the student-centred teaching method described in this report.

Notes

1. See, for example, Bainbridge (1980), Department of Education and Science (1978) and Plimmer (1981) commenting on the relative neglect of primary school science in Great Britain; De Rose, Lockard & Paldy (1979), Mechling & Oliver (1983), Neuman (1981), Perkes (1975) and Rudman (1978) making similar comments about the United States; and Alford & Kerrison (1974), Appleton (1977), Stamp (1982), and Symington (1974) describing Australian primary science likewise.
2. Now the School of Education, University of Waikato.

References

Alford, B. and Kerrison, R. (1974) Putting primary science into the primary school. *Australian Science Teachers Journal* 20, (1), 11–15.

Appleton, K. (1977) Is there a fairy godmother in the house? *Australian Science Teachers Journal* 23, (3), 37–42.

Appleton, K., Hawe, E., Biddulph, F. and Osborne, R. (1984) So you think the guide materials look good! *Research in Science Education* 14, 206–12.

Bainbridge, J. (1980) Tchirrip. . .tchichirrip. . .tseep: An alarm call for primary science. *School Science Review* 61 (217), 623–38.

Bell, B. F., Kirkwood, V. M. and Pearson, J. D. (1990) Learning in Science Project (Teacher Development): The Framework. *Working Paper 406*. Hamilton: Centre for Science and Mathematics Education Research, University of Waikato.

Bell, B. F. and Pearson, J. D. (1991) Learning in Science Project (Teacher Development): The Framework—1991. *Working Paper 410*. Hamilton, Centre for Science and Mathematics Education Research, University of Waikato.

Biddulph, F. (1987) Modelling alternative science teaching approaches for preservice teachers. *Working Paper No. 403*. Hamilton: Science Education Research Unit, University of Waikato.

——(1989) Children's questions: Their place in primary science education. *Unpublished Doctoral thesis*, University of Waikato, New Zealand.

——(1990) I still have a way to go: Case study in preservice teacher change. *Working Paper No. 407*. Hamilton: Centre for Science and Mathematics Education Research, University of Waikato.

——(1991) Pupil questioning as a teaching/learning strategy. *SAME papers 1990*, 60–73. Hamilton: Centre for Science and Mathematics Education Research, University of Waikato.

——(1992) *Investigating our World*. Auckland: Applecross.

Biddulph, F. and Osborne, R. (1984) *Making Sense of our World: An Interactive Teaching Approach*. Hamilton: Science Education Research Unit, University of Waikato.

Biddulph, F., Osborne, R. and Freyberg, P. (1983) Investigating learning in science at the primary school level. *Research in Science Education* 13, 223–32.

Biddulph, F., Symington, D. and Osborne, R. (1986) The place of children's questions in primary science education. *Research in Science and Technological Education* 4 (1), 77–88.

Crossland, R. W. (1972) An individual study of the Nuffield Foundation Primary Science Project. *School Science Review* 53, 628–38.

Department of Education and Science (1978) *Primary Education in England: A Survey by Her Majesty's Inspectors of Schools.* London: HMSO.

De Rose, J. V., Lockard, J. D. and Paldy, L. G. (1979) The teacher is the key: A report on three National Science Foundation studies. *Science and Children* 16 (17), 35–41.

Dillon, J. T. (1988) The remedial status of student questioning. *Journal of Curriculum Studies* 20 (3), 197–210.

Henry, J. A. (1976) Contemporary primary science curricula in the United Kingdom. *Australian Science Teachers Journal* 22 (1), 61–8.

Lucas, K. B. (1974) Notable developments in elementary school courses. *Australian Science Teachers Journal* 20 (1), 33–48.

Mechling, K. R. and Oliver, D. L. (1983) Who is killing your science program? *Science and Children* 21 (2), 15–18.

Neuman, D. B. (1981) Elementary school science for all children: An impossible dream or a reachable goal? *Science and Children* 18 (6), 4–5.

Perkes, V. A. (1975) Relationships between a teacher's background and sensed inadequacy to teach elementary science. *Journal of Research in Science Teaching* 12 (1), 85–8.

Plimmer, D. (1981) Science in the primary schools: What went wrong? *School Science Review* 62 (221), 644–7.

Rudman, H. C. (1978) Science teaching in the eighties—a case of benign neglect? *Science and Children* 16 (2), 7.

Science 5/13 (1972) *With Objectives in Mind.* London: Macdonald Educational.

Shymansky, J. A., Kyle, W. C. and Alport, J.M. (1976) How effective were the hands-on science programmes of yesterday? *Science and Children* 20 (3), 14–15.

Stamp, K. (1982) Primary science in the eighties. *Unicorn*, Bulletin of Australian College of Education.

Symington, D. (1974) Why so little primary science? *Australian Science Teachers Journal* 20 (1), 57–62.

Symington, D. (1980) Scientific problems seen by primary school pupils. *Unpublished Ph.D. thesis.* Monash University, Melbourne.

Wastnedge, E. R. (ed.) (1967) *Nuffield Junior Science: Teacher's Guide 1.* London: Collins.

Westbury, I. (1983) How can curriculum guides guide teaching? Introduction to the symposium. *Journal of Curriculum Studies* 15 (1), 1–13.

NOTES ON CONTRIBUTORS

Derek Bell has been involved in teacher education for ten years. He was a member of the Science Processes and Concept Exploration (SPACE) Project based at Liverpool University's Centre for Research in Primary Science and Technology. He is on the editorial board of the ASE's publication *Science Teacher Education* and also the ASET committee. He is Senior Lecturer in Professional Studies at the Liverpool Institute of Higher Education.

Fred Biddulph and **Frank Carr** both work in the Centre for Science and Mathematics Education Research at the University of Waikato, New Zealand. Dr Biddulph was the project officer for the *Learning in Science Project (Primary)* and Professor Carr directed the *LISP (Energy)* programme and the preliminary year of the *LISP (Teaching)* programme. He is also director of the new *Technology Education Project*. Both are widely consulted on science education policy in New Zealand and have published a number of papers.

Frank Curragh and **Patricia Diamond** are from The Queen's University of Belfast, where Dr Curragh is a Senior Lecturer and also Dean of the Faculty of Education. He is a member of many professional organisations, and is currently the Chairman of the Royal Society of Chemistry (Education Division—Ireland). He is a past manager of the University's Teachers' Centre and also Assistant Director of the School of Education's In-Service Division. Mrs Diamond is a Research Assistant working with Dr Curragh in the area of Primary Science Innovation and Effectiveness. Prior to this she taught and for some time she worked with the Belfast Education and Library Board, with particualr responsibility for probationary teachers.

Douglas Newton is a Lecturer in Science Education at the University of Newcastle upon Tyne, where he works on primary and secondary courses for both initial and inservice teachers. Dr Newton has published extensively on many aspects of science education and is particularly interested in ways of showing the relevance of science, an interest which led to his book, *Making Science Education Relevant* (Kogan Page, 1989).

Lynn Newton is a lecturer in Science Education at the University of Newcastle upon Tyne, where she is also Director of Initial Training (Primary). Mrs Newton coordinates the PGCE primary science course and also the 20-Day Science inservice course for primary teachers. Previously she taught in primary schools and was an advisory teacher for primary science.

Gill Nicholls is a Research Fellow and Senior Lecturer in Energy Education at Christ Church College, Canterbury. Within the college Dr Nicholls is the Course Director for the Licensed Teacher programme as well as teaching on PGCE and MA/MEd courses. She is currently leading a two year research project entitled *Energy 9–13* which is funded by British Gas.

Alan Peacock is a lecturer in primary science at the University of Exeter. After 11 years of teaching, Dr Peacock moved into teacher training both in the United Kingdom and in Kenya. He is a primary science consultant in Botswana, Namibia and India. His most recent book is *Science in Primary Schools—The Multicultural Dimension* (Macmillan, 1991).

Terry Russell is the Director of the Centre for Research in Primary Science and Technology at the University of Liverpool. He, along with his co-workers, Dr Derek Bell (now at the Liverpool Institute), Ms Linda McGuigan, Dr Anne Qualter, Mr John Quinn and Dr Mike Schilling, are all researchers at CRIPSaT. They have been involved in a range of evaluation, research and development projects centred on primary science and technology in the primary phase.

Robin Smith and **Graham Peacock** work at the Centre for Science Education of Sheffield City Polytechnic. Dr Smith has overall responsibility for primary science. Mr Peacock is responsible for the centre's in-service work. Individually and together they have produced films, books and other materials for teachers, their most recent being *SPECTRUM Science* (Collins, 1992) and *Teaching and Understanding Science* (Hodder and Stoughton, 1992).

Mike Summers and **Colin Kruger** work in the Department of Educational Studies at the University of Oxford, where Mr Summers is Director of the Primary School Teachers and Science (PSTS) Project. This project began in 1988 for four years and the research findings have already generated numerous papers and other publications. Both have published many papers on science education.

Catherine Woodward taught for eighteen years in a variety of schools across the full 4–18 age range. From a post as an advisory teacher for primary science she joined the Education Department at the University of Wales (Swansea) to teach science to primary PGCE students. She has worked with the Curriculum Council for Wales as a member of Steering and Task Groups which produced the Non-statutory Guidance (1980) and teacher support material.